BRITISH AND OTHER MARINE AND ESTUARINE OLIGOCHAETES

A NEW SERIES

Synopses of the British Fauna

No. 21

Edited by Doris M. Kermack and R. S. K. Barnes

BRITISH AND OTHER MARINE AND ESTUARINE OLIGOCHAETES

Keys and notes for the identification of the species

R. O. BRINKHURST

Ocean Ecology Laboratory, Institute of Ocean Sciences
P.O. Box 6000, Sidney, British Columbia, Canada

1982

Published for

The Linnean Society of London

and

The Estuarine and Brackish-water Sciences Association

by

Cambridge University Press

Cambridge

London New York New Rochelle

Melbourne Sydney

Published by the Press Syndicate of the University of Cambridge
The Pitt Building, Trumpington Street, Cambridge CB2 1RP
32 East 57th Street, New York, NY 10022, USA
296 Beaconsfield Parade, Middle Park, Melbourne 3206, Australia

First published 1982

Printed in Great Britain at the Pitman Press, Bath

Library of Congress catalogue card number: 81–3854

British Library Cataloguing in Publication Data

Brinkhurst, R. O.
British and other marine and estuarine
oligochaetes. – (Synopses of the British fauna.
New Series; no. 21)

1. Oligochaeta 2. Aquatic invertebrates
I. Title II. Linnean Society of London
III. Estuarine and Brackish-water Sciences Association IV. Series
595.1′46′0916 QL391.A6

ISBN 0 521 24258 4 hard covers
ISBN 0521 28559 3 paperbacks

Preface

Oligochaetes form one of the few groups which popular opinion debars from the sea. In fact the biomass of oligochaetes in intertidal stretches of various estuaries and sheltered coasts may exceed those of all the more-to-be-expected marine groups put together. Admittedly, their dominance has often coincided with the increase in organic pollution which typified many British estuaries since the early 1930s. Indeed, the author of this *Synopsis* has noted elsewhere that 'the relationship between organic pollution and the appearance of waving red carpets of worms has been recognized from antiquity, when the phenomenon was held to be due to spontaneous generation'.* This relationship has given a number of species the popular name of 'sludge worms'. Nevertheless, although they form the sole element of the macrofauna only in polluted estuarine areas, oligochaetes are also a natural part of the fauna of unpolluted marine shores, of subtidal zones and even of the deep sea.

The aquatic oligochaetes have the reputation of being a 'difficult' group and several marine and estuarine surveys have included a category entitled 'unidentified oligochaetes'. This is due in part to the extreme similarity externally between several species (one of the reasons why this *Synopsis* breaks with the tradition of illustrating each species as a whole animal) and in part to the lack of keys to the marine species, itself a reflection of our ignorance of the marine oligochaete fauna. There has been a great expansion in our knowledge of salt-water oligochaetes in the last ten years and much taxonomic uncertainty and confusion have been straightened out. Against this background, Dr Brinkhurst has produced this *Synopsis* of the British species and the world genera. It will permit the professional estuarine biologist to reduce considerably the category, mentioned above, of 'unidentified oligochaetes' in his surveys. We hope that it will also lead to a greater appreciation by both amateur and professional of this much neglected element of the British marine fauna.

R. S. K. Barnes
Estuarine and Brackish-water
Sciences Association

Doris M. Kermack
Linnean Society of
London

* Brinkhurst, R. O. (1980), Taxonomy, pollution and the sludge worm, *Mar. Poll. Bull.*, **11**, 248–51.

A Synopsis of the British and Other Marine and Estuarine Oligochaetes

R. O. BRINKHURST

Ocean Ecology Laboratory, Institute of Ocean Sciences
P.O. Box 6000, Sidney, British Columbia, Canada

Contents

Introduction

The marine and estuarine Oligochaeta belong to several families, mostly the Naididae, Enchytraeidae and Tubificidae, with the deep-sea forms largely restricted to the latter. All of these also have freshwater and even terrestrial relatives; although the true diversity of salt-water species is only now becoming fully appreciated. Many species have been described and many of the small, often monotypic, marine genera have been revised drastically since the aquatic oligochaetes were reviewed by Brinkhurst and Jamieson (1971). While this increase in our knowledge has clarified some of the subfamilies in the Tubificidae, it has not so far greatly affected the organization of the higher taxa provided in that review. The Oligochaeta were seen therein as a subclass of the Clitellata, this class also including the leeches (Hirudinea) and the Branchiobdellida (parasites on crayfish). In a forthcoming reference work on taxonomy and classification of living organisms (Brinkhurst, 1981), the Oligochaeta are regarded as a class in order to conform to the overall scheme adopted by the editor, but in either instance the classification at the level of orders remains virtually intact. Two of the former orders are monotypic, containing the families Lumbriculidae and Moniligastridae respectively, although it has recently been suggested that the latter is but a suborder of the Haplotaxida which includes all of the marine and estuarine species.

The Haplotaxida comprises at least three suborders. The first contains the stem family Haplotaxidae, many members of which occupy ground-water habitats that may be refugia or that may represent an ancestral pathway for the colonization of inland waters from a marine origin. The suborder Lumbricina has several aquatic stem-forms but for the most part constitutes the various families of earthworms. *Criodrilus lacuum* has been found in both fresh and brackish water in Europe, and *Pontodrilus* is a euryhaline megascolecid found in North America and Africa at least. Otherwise there are few if any members of this suborder that require consideration here, and so the text will be restricted to coverage of the third suborder, the Tubificina, amongst which the family Enchytraeidae is the only one not primarily aquatic; even that family is thought to have had an aquatic origin. There is one purely marine genus in the Enchytraeidae and marine or brackish-water species are located in other genera, but despite some recent parochial interest there has not been any recent taxonomic revision of the family, though there is a guide to European littoral forms (Tynen and Nurminen, 1969). Of the remaining families of the Tubificina, the Opistocystidae (revised by Harman and Loden, 1978) contains four or five freshwater species and the Phreodrilidae is a 'Gondwanaland' family with no European representatives. The Dorydrilidae, formerly regarded as part of the Lumbriculidae, consists of three European freshwater species with the possible addition of the poorly known genus *Lycodrilus* from Lake Baikal.

This leaves the Naididae, with several coastal and estuarine species, and the Tubificidae, which can be found from mountain springs down to the ocean abyss. Most of the species in these aquatic families are limited to freshwater, and the British freshwater fauna has been reviewed by the author in the *Scientific Publications of the Freshwater Biological Association* (FBA) (No. 22, 1963*c*). That publication should be due for revision shortly, at which time a few nomenclatural changes need to be made and some taxonomic decisions reviewed. The largest change required would be in the salt-water species, in which the *potential* British fauna list must now be considerably extended. When the FBA booklet was prepared, the salt-water species were few in number, and were included for convenience although they lay outside the coverage of the series. As many limnologists are already familiar with that work, and duplication has been deemed undesirable, the present work will cover only those species likely to be encountered to seaward of the heads of estuaries. The most resistant freshwater species will be included, as well as the genuine estuarine and marine forms. The faunal list from Britain forms an inadequate basis for consideration as so little work has been done there, and several species are cosmopolitan, so the approach taken here is to familiarize the student with the 'state of art' in salt-water oligochaete taxonomy by describing the most relevant species and by giving brief accounts of the diagnostic characteristics of related species. As there are so few species involved, an unusual style of keying will be employed that makes use of a series of decision levels. This device enables one to make decisions to a species-group level (not necessarily genus) based on a few readily visible characteristics observed on whole mounts made in a very simple manner. It also stabilizes the key in relation to the vagaries of generic names. It will not be possible to identify specimens under the stereo-microscope until and unless a detailed study of whole mounts, often with sections in marine forms, has narrowed the species list for a given habitat to a few recognizable forms.

The family Aeolosomatidae (which may be a higher taxon with several families here treated as genera) does not belong in the Oligochaeta. The only features they have in common (setae, segmentation, coelom, hermaphrodite reproductive system) are insufficient to place the group within the Oligochaeta, as these features are also present in aberrant polychaete families such as the Questidae for example, and the sequence of the gonads places them outside the otherwise monophyletic oligochaetes (Brinkhurst and Jamieson, 1971). The details of all the other anatomical features also differ from those of the oligochaetes (setae, nerve cord, septa, pharynx, prostomium, etc.) so that they should be regarded as a distinct class by themselves. Although there are some records of terrestrial species, most are from freshwater, but *Aeolosoma litorale* is found in brackish water (Bunke,

1967) and *A. maritimum* Westheide and Bunke is regarded as the first truly marine species in the family. *Potamodrilus fluviatilis* is found in interstitial spaces (the psammon) of rivers in Russia, Poland and Germany and on a Baltic beach, but is excluded here as an unlikely component of British beaches. The same treatment is accorded another aeolosomatid, *Rheomorpha neizvestnovae*, which has a very similar distribution, with the addition of the Alps. Both of these are, in fact, the sole representatives of their respective genera and have been regarded as having familial status. They are described in Brinkhurst and Jamieson (1971).

The major familial revisions pre-date the appearance of the world revision of aquatic oligochaetes by Brinkhurst and Jamieson (1971) in which more detailed treatment of various subjects may be found.

General biology

Morphology

External features

Like their terrestrial relatives, the earthworms, aquatic oligochaetes are segmented, bilaterally symmetrical, cylindrical animals with tapering ends (Fig. 1), but unlike several of the earthworms, most aquatic species are very small. Few are longer than 2 cm and most have diameters of only a fraction of a millimetre: usually they are some 30–60 times longer than wide. The body plan is a pre-segmental head or **prostomium**, followed by a series of more or less similar **segments**, and a post-segmental tail or **pygidium**: all cylindrical.

The prostomium and the first segment, the **peristomium**, together frame the ventrally directed, subterminal **mouth**; the pygidium bears the **anus**. The prostomium may have an elongate extension, the **proboscis**, and also simple pigment spots (termed '**eyes**') may occur in some though not all individuals of some naidid species.

Typically each segment except the peristomium possesses four **bundles** of **setae** (or chaetae*), two bundles dorsolaterally and two ventrolaterally. Each bundle may contain from one to twenty chitinous and proteinaceous setae of the same or of a variety of types and sizes. Sometimes the setae are missing from a few of the anterior dorsolateral bundles (as in the naidid *Paranais*), from a large number of segments (e.g. in the enchytraeid *Grania*) or even from all segments (in the enchytraeids *Achaeta, Marionina achaeta* and *M. arenaria*). Two basically different setal types can be recognized: the slender, elongate and straight **hair setae**; and the more diverse, stouter, often curved, **sigmoid setae** which may possess a swelling, the **nodulus**, where they emerge from the body wall. Hair setae are confined to the dorsolateral bundles of a few families including the Naididae and Tubificidae, whereas the sigmoid type are more widespread. Because of its importance in identification, setal diversity will be described in a separate section below.

As in other annelids, the body wall of oligochaetes comprises a thin **peritoneum** bounding the **coelomic body cavity**, an inner layer of **longitudinal** and an outer layer of **circular muscle**, an **epidermis** and an external, non-cellular **cuticle** secreted by the latter. The setae arise from epidermal **setal sacs** and have a variety of muscles attached to their bases enabling the setae to be moved during locomotion. The body wall is translucent and unpigmented in many species, permitting the internal organs to be seen

* In view of the spelling of the Oligochaeta and Polychaeta, and the use of the term seta (pl. setae) in other animal groups, it may seem logical to adopt 'chaetae' here rather than 'setae'. The latter, however, is the more frequently used of the alternatives in the recent oligochaete literature and is that favoured by most students of the oligochaetes.

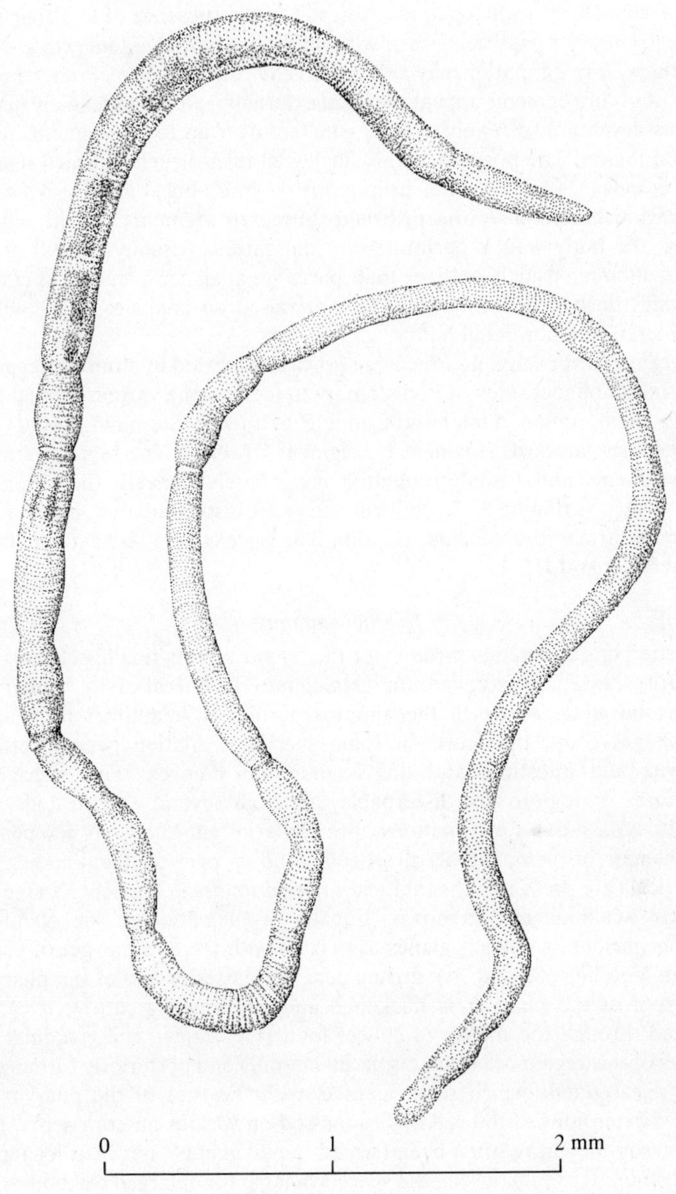

Fig. 1. A common coastal marine oligochaete *Tubificoides benedeni* (Udekem). From life; scale in mm.

through the skin, although some species appear red by virtue of possessing the respiratory pigment haemoglobin. In a few species, however, the body wall is characteristically opaque as a result of the presence of leaf-shaped or conical, cuticular **papillae**, with or without foreign particles adhering to them. In others, foreign matter may adhere directly to the cuticle.

Most of the segments appear identical externally, although sexually mature worms develop a relatively opaque **clitellum** over up to six segments in the genital region. The position of this single-cell-thick structure which secretes the **cocoons** (see p. 16) varies from group to group but it usually lies within the region bounded by the fifth and thirteenth segments. In the clitellar region the body wall is perforated by the various (usually paired) **genital pores**, although usually only the **male pores** are at all conspicuous, as circular openings flush with the body surface or raised on papillae. These will be considered in more detail below.

Because various organs are characteristically located in different segments in different oligochaetes, it is customary to identify the various segments by giving them roman numerals beginning at the peristomium (no. I) and proceedings tailwards (segment II, segment III, etc.). The septa separating the segments and (usually) situated immediately beneath the externally visible furrows ringing the cylindrical bodies are also numbered, but in arabic numerals to avoid confusion. Septum 2/3, for example, is that separating segments II and III.

Internal anatomy

The straight **gut** extends throughout the length of aquatic oligochaetes and comprises the subterminal mouth opening into the **buccal cavity**, a **pharynx**, a long **intestine** (of which the anterior portion is sometimes termed the **oesophagus**), and the anus. In some species a dilation occurs between pharynx and intestine, and this is then the **stomach**. Few specialized structures or regions are discernable, although several enchytraeids, e.g. *Henlea* spp., possess evaginations of the anterior gut known as **oesophageal appendages** or **oesophageal diverticula**, and a pair of blind-ending **gut diverticula** are present in the tubificid *Limnodriloides* in segment IX (Fig. 2*c*) – their functions are unknown. Equally problematic are the gland-like **peptonephridia** or **salivary glands** associated with the anterior gut of enchytraeids (see Figs. 35 and 36), arising near the posterior end of the pharynx. The roof of the pharynx is thickened and has a heavy cuticle: it can be everted through the mouth to collect food (Fig. 2*a*,*b*). The glandular cell bodies of this region are often displaced laterally and posteriorly forming the **pharyngeal glands** which retain contact with the roof of the pharynx via narrow extensions of the cell bodies massed on various anterior septa. (For this reason they have often been termed 'septal glands'; but their location is simply a matter of increasing the space available for enlarged cell bodies in a vermiform animal.) Some aquatic oligochaetes lack a gut, for example several species of *Phallodrilus*.

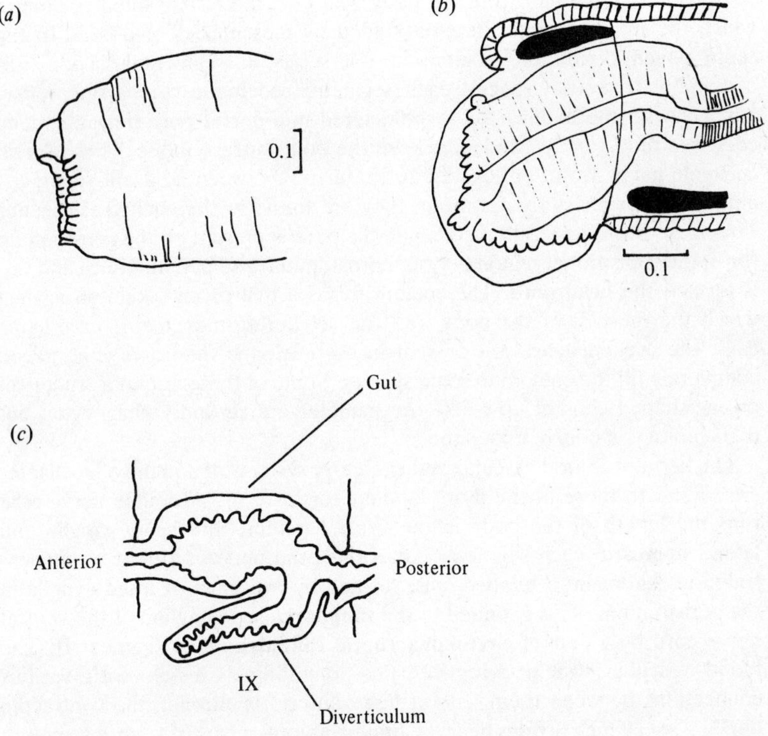

Fig. 2. Pharynx of *Ilyodrilus mastix* inverted (*a*) and everted (*b*). Scale in mm (after Brinkhurst, 1978). (*c*) Gut diverticulum in *Limnodriloides* (after Hrabe, 1971*a*).

The gut is separated from the body wall by a large fluid-filled coelom in which the internal organs are suspended by mesenteries or affixed to the septa which divide the coelom into a series of segmental cavities. In essentially terrestrial species each segmental coelomic compartment may connect with the exterior by a sphinctered mid-dorsal pore through which coelomic fluid can be exuded to keep the body surface moist. These **dorsal coelomic pores** are often located in the furrow between adjacent segments; amongst the species treated here, they are found in the enchytraeid genus *Fridericia*. More generally, a comparable pore is present on the prostomium (or in the transitional region between prostomium and peristomium) and this is termed the **head pore**. The coelom forms a hydrostatic skeleton against which the muscles of the body wall can act during movement. Free-living cells, the **coelomocytes**, are present in the coelomic fluid and may indeed completely fill the coelom in some species. Some of these granular, spherical or egg-shaped cells of up to 20 μm diameter are certainly phagocytic, but their biology is poorly understood.

The nervous, blood vascular and excretory systems of aquatic oligochaetes are similar to those of the more familiar earthworms. A double **nerve cord** runs the length of the body in the ventral midline and bears **ganglia** and lateral nerves in each segment. Anteriorly, the nervous system is concentrated into a **brain**; a bilobed mass formed by two pairs of fused ganglia in the peristomium. This is joined to the **subpharyngeal ganglion** of the ventral nerve cord by a pair of **circumpharyngeal commissures** in segment II. The blood vascular system comprises two longitudinal vessels with various connections between them. Blood flows anteriorly through the contractile dorsal vessel which divides near the brain and joins a pair of ventral vessels in the peristomium. These two merge somewhere between segments III and VII to form a single ventral vessel lying just dorsal to the nerve cords, through which blood flows posteriorly. Flow from the ventral to the dorsal vessel in each segment is through a plexus of vessels around the gut, and in the reverse direction is through the circumintestinal connectives. Gaseous exchange is simply by diffusion across the body surface although some freshwater species do possess **gills** – thin-walled extensions of the body wall. Most aquatic oligochaetes lack respiratory pigments, but when present – as in several tubificids – they are in the form of haemoglobin dissolved in the blood plasma.

Excretion and osmoregulation are achieved by **nephridia**. These open-ended tubes take in coelomic fluid through a **ciliated funnel**, pass it along an elongate **nephridial tube** (in which there is selective reabsorption) and discharge it to the exterior through a small pore, the **nephridiopore**, near the ventral bundle of setae. One pair of nephridia is typically present in each segment except for a few anterior and genital ones which lack them. Each nephridium penetrates a septum so that the ciliated funnel is sited anteriorly to that septum (forming the **anteseptale** of the nephridium) whilst the

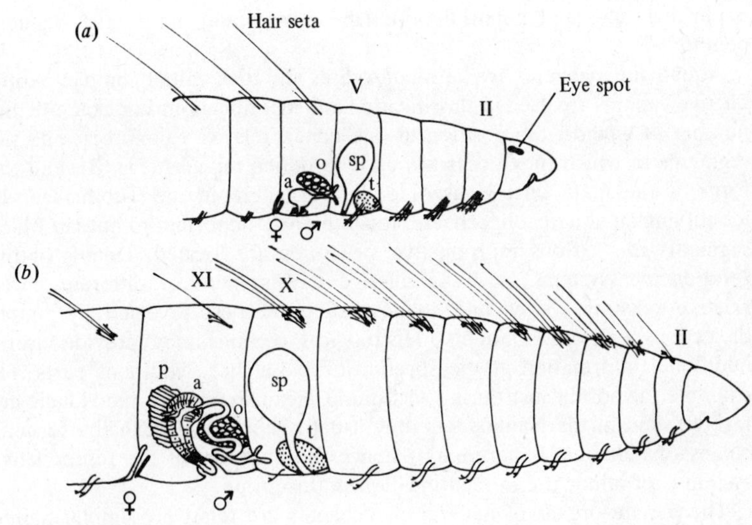

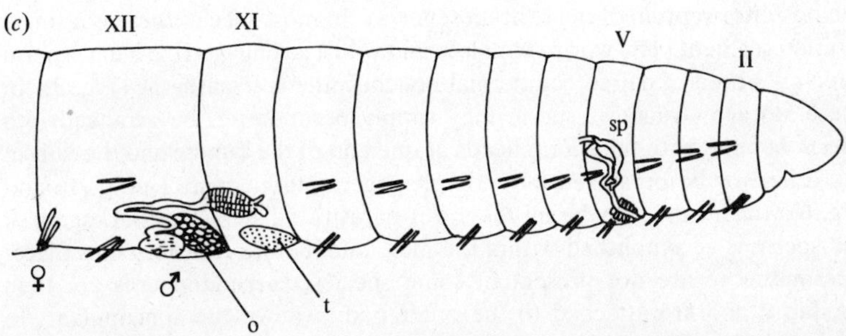

Fig. 3. Arrangement of setae and reproductive system in the Naididae (*a*), Tubificidae (*b*), and Enchytraeidae (*c*). All reproductive organs are paired, although for clarity only one of each is shown. Note that the segmental position may be further forward or backward in some species. Key: a = atrium; p = prostate gland; o= ovary; t = testis; sp = spermatheca; ♀, =♂ = genital pores.

nephridial tube and nephridiopore (the **postseptale**) are in the segment behind.

Aquatic oligochaetes are hermaphrodites and their rather complex reproductive systems are used in classification and identification. For example, the number of **gonads**, the position of one gonad relative to another, and the segments in which they occur are used to define families (Fig. 3); and the form of the male duct is used to define genera in the Tubificidae. In identifying an aquatic oligochaete, it is therefore important to note in which segments the various reproductive structures are located. Details of the reproductive systems, such as relative development of different parts, relative positions, sizes and shapes, etc., are best revealed by simple dissections or whole mounts. Histological sections may provide useful additional information at the specific level, but hard chitinous parts, for example, are often lost during sectioning. Sexually mature individuals are rarely found in the Naididae as they usually reproduce asexually, forming chains of individuals (**paratomy**); hence the anatomy of the reproductive system is of minor use in identification in this group.

The female organs of aquatic oligochaetes are relatively simple: paired **ovaries** are attached to the anterior septum of the ovarian segment and the eggs are discharged to the exterior through paired, ciliated **female funnels** on the posterior septum of the ovarian segment. In most species there is a single ovarian segment. The worms copulate and whilst so engaged exchange sperm which is stored in paired ectodermal pouches, the **spermathecae** (Fig. 3). In these storage structures sperm may simply occur loose, be arranged into **sperm bundles** with the sperm heads at one end of the bundle and the tails at the other, or be organized into complex **spermatozeugmata** (see p. 16 and Fig. 6) which can easily be mistaken for parasitic ciliates. This packaging of the sperm is accomplished within the male duct before transfer takes place. Spermathecae are not present in some species: **spermatophores** are then produced and are attached to the outer body wall of the concopulant. In enchytraeids the spermathecae comprise a sac-like **ampulla**, often with diverticula, and sometimes with a duct connecting the ampulla to the oesophagus.

The male organs, which are located in segments anterior to those bearing the ovaries, comprise one or two pairs of **testes** (usually one pair but rarely more in parthenogenetic forms) and their associated complex male ducts (Fig. 4). The testes are sited on the anterior septa of the testicular segments and produce sperm which may be stored in **sperm sacs** or **seminal vesicles** on the septa whilst undergoing the later stages of spermatogenesis. A pair of ciliated **male funnels** or **sperm funnels** on the posterior septum receive the mature sperm and pass it into ciliated **vasa deferentia** (singular **vas deferens**) which themselves discharge into paired or single **atria** with a presumed storage function in most families. The vasa deferentia pass through the septum so that the atria or atrium occur in the segment posterior to that bearing the testes. Aggregations of gland cells, known as the **prostate glands**,

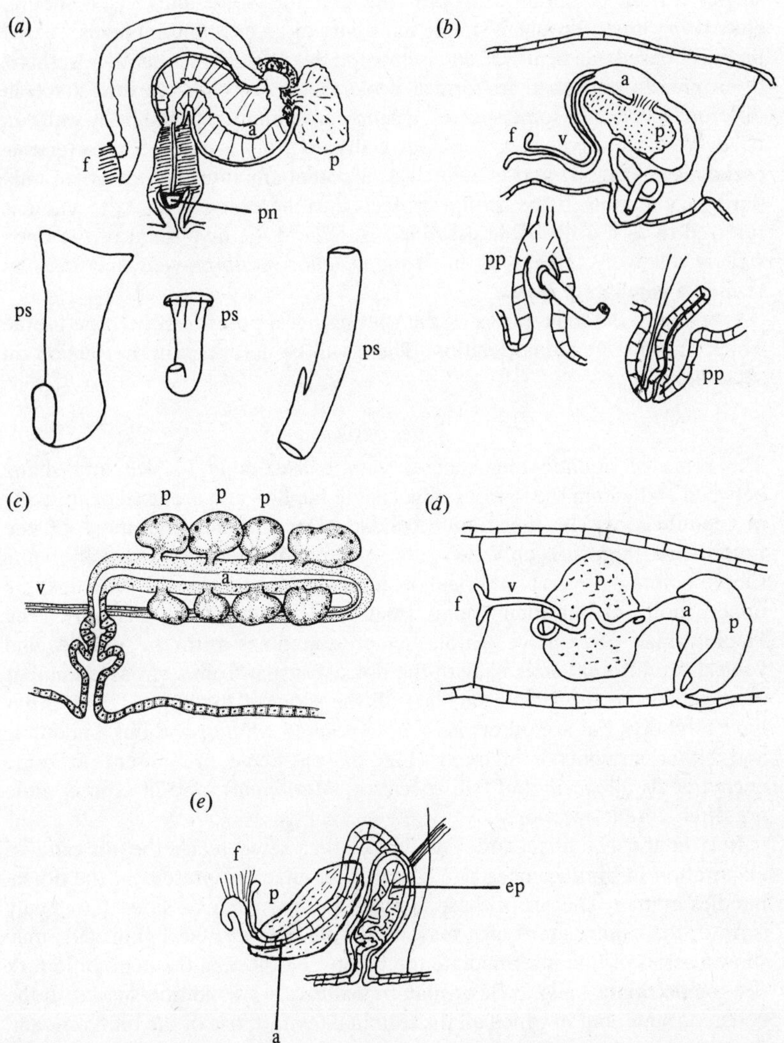

Fig. 4. Arrangements of the male reproductive system in the various tubificid subfamilies: (*a*) Tubificinae; (*b*) Aulodrilinae; (*c*) Telmatodrilinae; (*d*) Phallodrilinae; and (*e*) Rhyacodrilinae. Key: a = atrium; ep = eversible pseudopenis; v = vas deferens; p = prostate; pp = pseudopenes, retracted and protruded; pn = penis; ps = cuticular penis sheath; f = funnel.

are associated with the usually muscular atrium and each (or, rarely, the single) atrium communicates with the exterior, sometimes via a tubular **ejaculatory duct**, through a simple **male pore** or an intromittent **penis**. A true penis is housed in a **penis sac** and is covered by cuticle: in the tubificids, this is often greatly thickened to form a **penis sheath**, of characteristic shape in different species. In some species, a simple invagination of the body wall can be everted to serve as a copulatory organ. This is termed an **eversible pseudopenis** (Fig. 4) and is everted as a potentially intromittent organ only during copulation. If the atrium projects into the eversible sac, the whole is referred to as a **protrusible pseudopenis**. Where the involution of the body wall is relatively large – as in *Monopylephorus rubroniveus* – it may be termed a **copulatory bursa**.

Frequently the male pores or the spermathecal pores bear modified setae which presumably aid copulation. These will be described in the section on setae below.

Types of setae

The setae of aquatic oligochaetes vary considerably in size and shape between and within the various oligochaete families and are used extensively in identification. In the Lumbriculidae, Dorydrilidae and most of the Lumbricina there are only two setae per bundle, each seta usually being simple-pointed. Several lumbriculids, however, have setae with bifid tips, the upper tooth of the pair being small and rudimentary (Fig. 5*j*). The Haplotaxidae often have simple setae, sometimes with the dorsals and ventrals of different sizes or with the dorsals absent from a variable number of segments. In the Enchytraeidae, all the setae of the aquatic *Propappus* have bifid tips, but are otherwise simple-pointed, with or without a nodulus, and either straight or sigmoid (Fig. 5*k–n*). Setae are absent in some enchytraeids (*Achaeta* and two species of *Marionina*), and in *Grania* setae are often absent anteriorly.

It is in the Naididae and Tubificidae that setae reach the ultimate of elaboration in both number and form. Hair setae are present in the dorsal bundles in many species of these families and they may be smooth or finely serrate; the other setal types may be simple-pointed, bifid (Fig. 5*e*), may show a series of fine intermediate teeth between those of the main bifid fork (i.e. be **pectinate** – Fig. 5*d*), or may be **palmate** – a condition similar to the pectinate state, but in which all the terminal teeth (those of the bifid fork and of the intermediates) are of equal size. In the naidids the dorsal setae (other than the hairs) are termed **needles**; they are usually thinner and straighter than the ventrals and have smaller teeth. This feature aids in the separation of naidids from immature tubificids.

This elaboration of setal types in some aquatic oligochaetes (in relation to the terrestrial species for example) might seem to parallel the setal diversity of the thoroughly aquatic polychaetes. However, it is difficult to understand

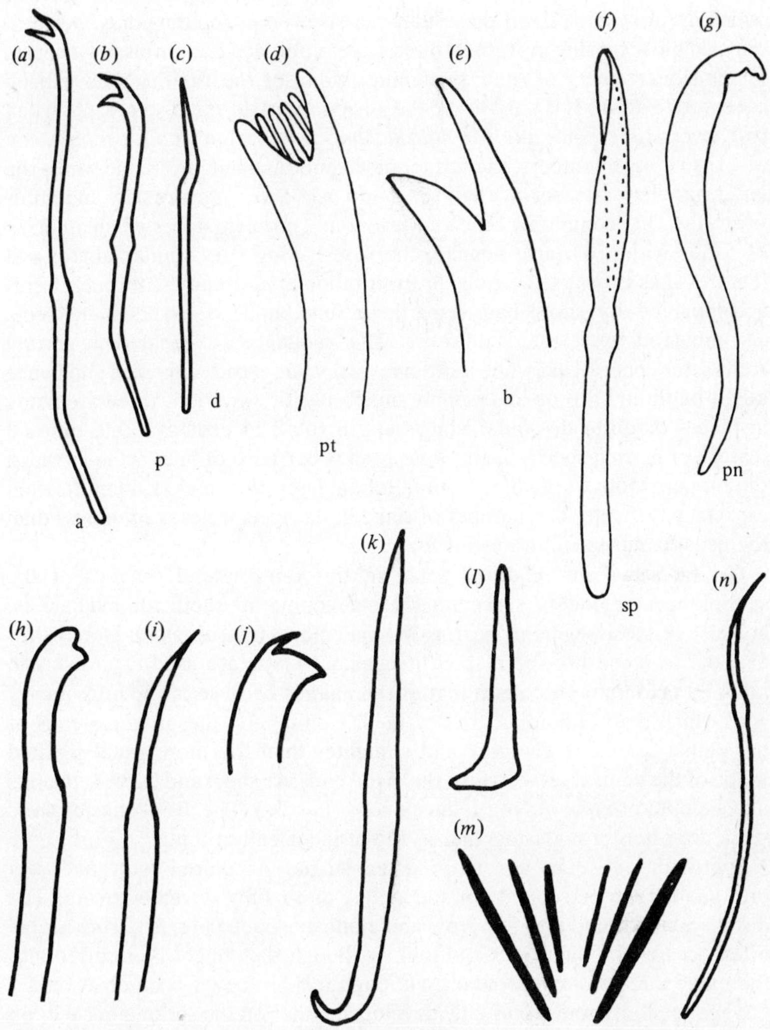

Fig. 5. Oligochaete setae: (*a*)–(*c*) Naididae; (*d*)–(*g*) Tubificidae; (*h*)–(*j*) Lumbriculidae; (*k*)–(*n*) Enchytraeidae (*k, Lumbricillus, l, Grania, m, Fridericia, n, Cernosvitoviella*). Key: d = dorsal; a = anterior ventral; p = posterior ventral; b = bifid (present in some form in all ventral bundles of tubificids, and in the dorsals when both hair and pectinate setae absent); pt = pectinate (usually with hair setae in dorsal bundles in tubificids, rarely in naidids); sp = spermathecal (usually the ventrals of X); pn = penials (usually the ventrals of XI) (spermathecals and penials, if present, are the genital setae of mature individuals).

the function of the wide range of setal types in the oligochaetes, most of which do not swim. Even those that can swim, e.g. some naidids, perform only simple spiralling movements and never approach the errant polychaetes in the effectiveness of their swimming. None of the tubificids have been observed to swim. It is hard to see the advantage of different setal types when two successful species like *Limnodrilus hoffmeisteri* and *Tubifex tubifex* can co-exist in huge numbers, the former having nothing but bifid setae whilst the latter has elaborate, serrate hair setae and pectinate setae dorsally and bifids ventrally. One commonly repeated taxonomic problem arises when an array of forms within a single genus, otherwise having very similar anatomical features, has dorsal setae ranging from bifids in all bundles, through bifids accompanied by short hair setae in a few bundles, to pectinate setae accompanied by serrate hair setae. This sequence is repeated in several freshwater genera, and while it often involves sympatric congeners and hence might be thought to be an example of subspecific variation, there are many instances of quite dissimilar congeners in tubificid ecology. One unusual feature of marine species is the widespred occurrence of bifid setae in which the upper tooth is reduced (in all bundles); this makes identification especially difficult. The number of setae in marine species is often less than the usual count in frehwater forms.

Genital setae are modified setae in the ventrolateral bundles of the spermathecal segments (**spermathecal setae**) of mature tubificids and naidids, and also of those segments bearing the male pores (**penial setae**). Rarely they also occur in the pre-spermathecal segment. These are used extensively in tubificid taxonomy, less so in that of the naidids because of the infrequency with which mature naidids are obtained. Commonly, the proximal ends of the penial setae are elongate and straighter than the more usual sigmoid shape of the ventral setae, whilst the distal ends are short and curved; the tips may be bifid or enlarged to produce knobs (Fig. 5*g*). The distal ends of these setae are often brought together as in the inner end of a fan.

Spermathecal setae are more often single, or paired with one seta presumably a replacement for the other, more fully developed seta. The distal end is formed into a narrow and elongate spoon (Fig. 5*f*). These setae often occupy glandular sacs and look as though they might be inserted into the male pores of a concopulant, although this has never been observed.

These genital setae are usually developed either on the spermathecal or on the penial segment, rarely on both; they are usually absent in those species which possess penes with cuticular sheaths. Various combinations of these aids to copulation can be found, however, and all three are developed in one unusual freshwater tubificid.

(The Aeolosomatidae possess similar setae in both dorsal and ventral bundles (in *Aeolosoma* and *Potamodrilus*), although setae are absent in *Rheomorpha* and the setae of the dorsal and ventral bundles do differ slightly in the parasitic genus *Hystricosoma*.)

Feeding

Very few studies on feeding of aquatic oligochaetes have been carried out. The naidid *Chaetogaster* is at least partially a predator (Green, 1954; Poddubnaja, 1965), utilizing a wide variety of prey, but many other naidids utilize bacteria, or are probably herbivores or utilize the 'aufwuchs' or epiphytic flora (Learner, Lochhead and Hughes, 1978). The tubificids ingest sediment, almost certainly deriving the bulk of their nutrition from bacteria (Brinkhurst and Chua, 1969; Wavre and Brinkhurst, 1971) and perhaps from algae (Moore, 1978). That worms mediate passage of heavy metals from sediment via bacteria to fish has been demonstrated by Patrick and Loutit (1976, 1978). Several predators make use of aquatic oligochaetes, including leeches (Cross, 1976), fish (Kennedy, 1969; Aarefjord, Borgstrom, Lien and Milbrink, 1973), ducks (Rofritz, 1977) and a variety of invertebrates such as chironomids (Loden, 1974).

The clean-up of the Thames below London led to much improved oxygen levels but left so much organic sediment on the bottom that tubificids thrived, at least for a while, and they were exploited by the large flocks of various kinds of birds that moved into the estuary (Harrison and Grant, 1976), putting sludge worms in the public eye.

The disturbance of sediments by the feeding activities of tubificids and their effect on chemical balances therein have been studied by Davis (1974*a*,*b*), Davis, Thurlow and Brewster (1975), Wood and Chua (1973, 1977), Whitten and Goodnight (1969), McCall and Fisher (1980); the uptake of metals by Chapman, Churchland, Thomson and Michnowsky (1980).

Suction-feeding in the Aeolosomatidae was described by Singer (1978), who demonstrated that these worms utilize highly degraded plant material with its associated microbiota.

Enchytraeids in terrestrial habitats select plant remains rich in fungal material and are thought to feed upon the fungi and bacteria involved in the decomposition of plant material (Dash and Cragg, 1972). Direct uptake of amino acids via the integument was illustrated by Brinkhurst and Chua (1969) in tubificids, and has since been quantified by Siebers and Ehlers (1978) in *Enchytraeus albidus*. The discovery of marine oligochaetes with no alimentary canal suggests that this ability may become enhanced in some coral-reef dwelling tubificids (Erseus, 1980*c*).

Life history and ecology

Aquatic oligochaetes frequently reproduce asexually, predominantly so in the Naididae, for example, where chains of individuals may frequently be observed. The Tubificidae reproduce by simple fission, in laboratory cultures at least. In the sexual condition the worms are hermaphrodites and they copulate, though this has never been observed as it has been in earthworms.

The gametes mature in the coelom in extensions of the gonadial segments simply called egg or sperm sacs. The sperm enter the male ducts through the ciliated funnels, and may be formed into sperm bundles with the sperm heads at one end, or into spermatozeugmata with the heads around the inner core (Fig. 6), presumably within the atria (see p. 10). Copulation leads to the deposition of sperm in one form or another into the spermathecae of the concopulant. Self-fertilization is also a possibility. The sperm is liberated to fertilize small clutches of eggs shed into cocoons, in all probability secreted by the thin clitellum of these aquatic species. The secretions of the prostate glands are thought to nourish the sperm either before copulation or in the cocoon or both. Worms may produce several cocoons, but the five or six eggs in each may not all develop into the small replicas of the adult that are the ultimate product of development, as some may end up as food for the earliest to develop. Small worms with a few setae can be seen moving actively within the cocoons, which are found loose in the sediments.

Reproduction may be seasonal (Brinkhurst, 1964, 1966*a*, Cook, 1969*a*) or may be spread throughout the year and/or occur at different times in different locations (Kennedy, 1966*a*,*b*). Many worms seem to take two years to come to full maturity, so that in any given population there may be many immature worms at all times, but reasonable numbers of mature forms are usually present. Breeding worms contain sperm in the spermathecae, and the sperm may remain there for at least a month. The subsequent appearance of cocoons, followed by large numbers of very small individuals within weeks, can demonstrate breeding activity, but otherwise the population cannot be divided into age classes. In the laboratory, post-breeding worms can lose the reproductive ducts and subsequently develop a second set, but it is not known if this happens in the field.

Ladle (1971) set up cultures with cocoons of *Tubifex tubifex* at 16 °C in late February and these had produced mature worms from new hatchlings by mid June. By the beginning of September the mature forms were dead or regressed, and immatures dominated the cultures. Aston (1968) had earlier documented a simple annual life history for *Branchiura sowerbyi* in the field; whilst Thorhauge (1976) showed that breeding *Potamothrix hammoniensis* were four, sometimes three, years old, with total life spans of five years. Potter and Learner (1974) suggested that *Limnodrilus hoffmeisteri* could produce four or five generations a year in a small Welsh reservoir with temperatures between 17 and 18.6 °C for four months, whereas Ladle (1971) believed the same species produced but a single generation. Aston (1973) produced mature forms from eggs in five weeks at the elevated temperature of 25 °C, and while this rate is unlikely to be achieved in the temperate zone (outside locations affected by thermal discharges), the frequency of field sampling could determine the apparent life histories observed if, in fact, it is possible for several generations to be produced in a single season. Poddubnaja and Bonomi have made much more intensive studies of life histories (Bonomi and DiCola, 1980; Poddubnaja, 1980).

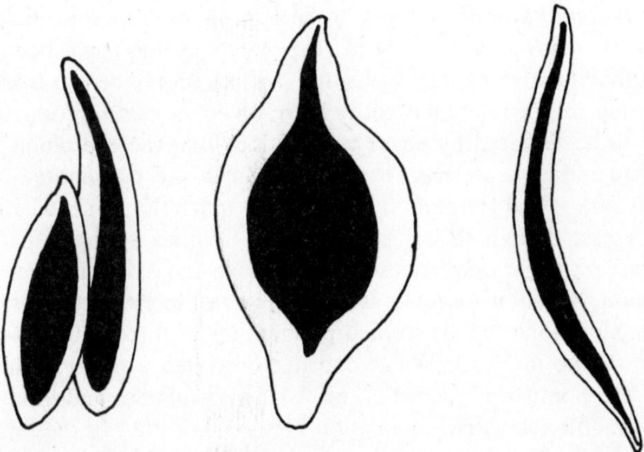

Fig. 6. Spermatozeugmata of tubificine tubificids. The sperm heads line the central (shaded) portion; the sperm tails are arranged spirally in the outer (unshaded) layer.

There have been few studies of truly marine species since that of *Tubifex costatus* (Brinkhurst, 1964). Hunter and Arthur (1978) suggested that *Tubificoides benedeni* does not breed in the first year of life in the Thames estuary, where they studied the ecology and pollution tolerance of this common species which has also been investigated by Wharfe (1977) in the adjacent Medway. Birtwell and Arthur (1980) document more recent Thames studies. Since the early studies by Knöllner (1935) and Bulow (1955, 1957), a series of publications by Arlt (1969), Dzwillo (1966), Giere (from 1970 to 1980) and Pfannkuche (1977) have extended the German contribution. Many of the papers by Erseus and Lasserre, either separately or as co-authors, are taxonomic and will be referred to below, but most include at least habitat information and zoogeographical data. The same is true for most of the works of Cook (Cook (1971) having a considerable amount of ecological information), Finogenova and Hrabe (both being concerned with the Black Sea, Caspian Sea and the Adriatic). Pollution studies in brackish areas are popular in Finland (Bagge, 1969; Leppakowski, 1975). Physiological work on salinity and respiration has been carried out, especially by Lasserre (reviewed in Lasserre, 1976), Giere (from 1970 to 1980) and Tynen (1969). This brief review is not meant to be exhaustive, but should enable readers to trace most of the literature from these initial sources.

Reproduction in the Naididae is reviewed by Learner *et al.* (1978). Although most naidids breed asexually by budding, *Allonais* fragments. Many naidids are most abundant in summer, stimulated by high temperatures and a plentiful food supply, but *Nais communis* buds more actively in cold weather, while *N. variabilis, N. obtusa* and *N. pseudobtusa* display little seasonal difference. In other instances conflicting evidence is reported. The estuarine *Paranais litoralis* was said to be a spring-time species. Species may disappear for many months only to reappear, and they have been held to burrow, inhabit resistant stages of some sort, or simply be too scarce to be collected, but this problem has yet to be resolved. Sexual reproduction does not seem to be initiated by adverse conditions, but these prohibit budding and so lead initially, at least, to a reduction in the population. There is apparently one sexual generation a year, after which the adults die or the sex organs regress, though this is not universal. Within a single species, the sexual phase may also vary from season to season and so in all respects sexual reproduction, when it occurs, seems to be as unpredictable as in the tubificids. McElhone (1978) studied populations of three freshwater naidid species in a lake in North Wales and demonstrated a positive correlation between the population densities of *Nais pseudobtusa* and *Chaetogaster diastrophus*, and a negative correlation between densities of the former and *Stylaria lacustris*. Population size and the proportion of individuals with buds could be correlated with temperature and/or food.

There have been several studies of the benthos of British estuaries in recent years, in which the larger oligochaetes, at least, have been identified. A few recent examples will serve to describe some typical results. Gray (1976) described the fauna of the Tees, an estuary classified as grossly polluted. *Peloscolex benedeni* and *Tubifex pseudogaster* (now both in *Tubificoides*) occurred in stable sand in the Seal Sand area, whereas *Paranais litoralis* seemed to prefer unstable sands; oligochaetes in general preferred rich organic sediments. *Clitellio arenarius* was also found, contributing to the high numbers of oligochaetes present (206 551 m^{-2}) which the author suggested were often overlooked in earlier studies because they were sieved away by benthic biologists and were too large to be considered part of the meiofauna. The oligochaetes and polychaetes seem to be a food source for the large, diverse bird population of the area. The same area was studied by Kendall (1979). In this seasonal study, *T. benedeni* reached its population maximum in late summer, and *Paranais litoralis* showed even stronger seasonal cycles. Populations of *Lumbricillus lineatus* differed at the two sites studied; at one the populations remained more or less constant, at the other they varied much like that of *T. benedeni*.

Wharfe (1977) described the Medway in Kent, and noted that *Tubificoides benedeni* increased in abundance, and the polychaete *Capitella capitata* appeared, when algal blooms decayed and oxygen levels were reduced. Gage (1974) also found *T. benedeni* in Loch Etive and Lake Creran, and recorded *Aktedrilus* (as *Phallodrilus*) *monospermathecus*, though in view of the number of species known with single median spermathecae, the identification should be regarded with some caution. *T. benedeni* was found in a wide variety of sediments.

In the Thames, Birtwell and Arthur (1980) found *Tubifex tubifex, Limnodrilus hoffmeisteri, Tubifex costatus* and *Tubificoides* (as *Peloscolex*) *benedeni* to be sequentially the most abundant species along the estuary from the freshwater to the marine end. Salinity is a primary factor, with dissolved oxygen, in determining distribution, with particle size setting population limits. The studies of Arthur, Huddart, Hunter and Birtwell are summarized in this report, the only other published account being that of Hunter and Arthur (1978) on *T. benedeni.* All four species were studied in relation to their tolerances to salinity, temperature, anaerobic conditions and sediment-type variations, and their respiratory physiology was studied in relation to their tolerance of low oxygen conditions. The life histories of *T. costatus* and *T. benedeni* were discussed in these two publications. The only other species mentioned is *Monopylephorus rubroniveus* at Erith Beach.

The Forth estuary is also heavily industrialized, the oligochaete fauna of the most polluted sections being detailed by McLusky, Teare and Phizachlea (1981). Once again, the upper estuary (interstitial salinity 0.2–4.1 ‰) is inhabited solely by *Tubifex tubifex* and *L. hoffmeisteri*, where the mean number at the most densely populated site reached 127 400 m^{-2}. At sites 16–28 km below Stirling, *T. costatus* and *Tubificoides benedeni* were dominant (mean interstitial salinities 3.2 ‰), the former reaching 444 933 m^{-2} or in excess of 78 g dry wt m^{-2}. The latter became more abundant beyond 30 km from Stirling (at which point *T. costatus* was absent), reaching 10 000 m^{-2} at 36 km where the mean salinity was in excess of 26 ‰ in the sediment. *Amphichaeta sannio, Paranais litoralis* and *Tubificoides pseudogaster* were also found, together with *Lumbricillus lineatus*, between Kincardine Bridge and Bo'ness (28–38 km below Stirling).

The distribution of the major species accords well with the Thames work and the importance of these dense worm populations as food for aquatic birds was emphasized (D. S. McLusky, personal communication).

Anne Henderson (personal communication) has found *Tubifex tubifex, T. costatus, Tubificoides pseudogaster, T. benedeni, Monopylephorus rubroniveus, Limnodrilus hoffmeisteri, Clitellio arenarius, Aktedrilus monospermathecus, Paranais litoralis, Nais elinguis, Stylaria lacustris* and *Lumbricillus lineatus* in the Clyde estuary, and some of these in other Scottish localities, plus *Monopylephorus irroratus* in the mouth of the River Cart, Renfrewshire, though this identification requires confirmation in view of recent studies.

Collection and preservation

Aquatic oligochaetes may be collected by a wide variety of procedures, depending upon the physical limitations of the situation. Many are associated with soft sediments, and so all of the coring and grab-sampling methods used by limnologists (Brinkhurst, 1974) or oceanographers (Holme and McIntyre, 1971) can be employed. Samples may be preserved *in toto* in the field, using formalin. Worms may be preserved immediately, or after sorting from preserved samples, in 70% alcohol, remembering to allow for the fluid content of the worms and water carried over. Many collections have reached my desk reduced to a suspension of empty cuticles because the alcohol became too dilute. Special methods obviously have to be employed for meiofauna and for delicate species of such genera as *Aeolosoma*. If the smaller species are not to be lost, screening should be minimal, with the smallest sieve sizes practical, especially in marine situations where the tubificids are unusually small.

Preparation for identification depends upon the individual needs, and while various claims to exclusivity are made, a variety of methods may have to be employed. For routine examination of large numbers of specimens from localities with a well-known fauna, especially if made up of a limited species diversity, simple temporary whole mounts in Ammans lactophenol are excellent, except for the Enchytraeidae. I place several worms under each cover-slip, two sets per slide, with the slides arranged on large trays. The worms may need a day or so to clear, and often require gentle pressure on the cover-slip to flatten them slightly. These preparations allow the setae and penis sheaths (if any) to be seen, and the reproductive structures of many tubificids may be visible using this method. The smaller, marine tubificids often require more detailed examination because the setal diversity is low and the male organs are often hard to discern without staining and/or some further manipulation. Whole mounts may be made in one of the 'instant' specifics, often with the benefit of an added dye, or they may be made more reliably permanent in the classical dehydrated balsam mount. Small worms may be halved sagitally, and mounted as two pieces (the head and tail could be mounted under a second cover so as to allow the halves of the reproductive segments to be gently flattened).

The male duct is best seen in its entirety in a dissection. The head and tail are removed and set aside, the genital segments are placed on the slide in the final mounting medium and are dissected *in situ* to avoid loss of vital fragments. The body wall is torn open mid-dorsally, folded aside, the gut is removed from back to front, and the male ducts teased off the body wall. The sperm funnels are often torn off with the gut, and the penes may be left on the body wall unless carefully teased free of it. With practice, this method gives a great deal of useful information quite fast, and is needed where new species are obtained or confirmation of identity is sought. Dissecting in the final medium makes it difficult, but not impossible, to use balsam, and so some preparations suffer from the shortcomings of the instant media which may break down into bubbles, even if the cover is sealed (with 'glyceel', a ringing material supplied by Gurr Ltd, for instance, or coloured nail polish that can also be used as a coding system). However, these media are usually soluble, and can be simply replaced in older slides, and the retrieved pieces may even be re-arranged.

Sections are often needed to answer specific questions, most often related to the form and placement of the prostate glands. Reconstruction of the male ducts (and especially the penis sheaths) from sections without dissections is sometimes advocated, but has its limitations and is time consuming. The illustration of the male ducts in new species from sections without the benefit of a sketch of the reconstructed whole is one of the more frustrating aspects of many recent papers. The form and proportion of the various component parts of the male duct are the most important generic characters in the tubificids, and as such a drawing of a dissection is the single most useful component of a description. With rare specimens, I much prefer to dissect (or use a whole mount if the worm is very small) rather than risk the sectioning process and the interpretation involved.

The reference collection may be stored on slides, or in vials (especially homeopathic lipped vials with neoprene stoppers) which do not allow alcohol to evaporate. Glycerine must not be added to the preservative. Brittle specimens are said to be softened by immersion in liquid detergent, but I cannot guarantee this.

Classification*

Phylum ANNELIDA
Class OLIGOCHAETA
Order HAPLOTAXIDA
Suborder TUBIFICINA

Family NAIDIDAE
  Subfamily CHAETOGASTRINAE
    Chaetogaster langi Bretscher
    Chaetogaster diaphanus (Gruithuisen)
    Chaetogaster crystallinus Vejdovsky (?)
    Amphichaeta sannio Kallstenius
  Subfamily PARANAIDINAE
    Paranais litoralis (Müller)
    +*Paranais frici* Hrabe
  Subfamily NAIDINAE
    Uncinais uncinata (Orsted)
    Nais communis Piguet
    Nais variabilis Piguet
    Nais elinguis Müller
    Stylaria lacustris (Linnaeus)
Family TUBIFICIDAE
  Subfamily TUBIFICINAE
    Tubifex tubifex (Müller)
    Tubifex nerthus Michaelsen
    Tubifex costatus (Claparede)
    +*Tubifex litoralis* (Erseus)
    Tubificoides amplivasatus (Erseus)
    Tubificoides pseudogaster (Dahl)
    +*Tubificoides aculeatus* (Cook)
    +*Tubificoides heterochaetus* (Michaelsen)
    Tubificoides benedeni (Udekem)
    Limnodrilus hoffmeisteri Claparede
    Isochaetides michaelseni (Lastockin)
    Clitellio arenarius (Müller)

* This list contains euryhaline and marine species that are or are likely to be found in Britain, plus all the known marine genera as all of them may be found in offshore locations. Species marked + are not as yet on the British list. List last revised May. 1981.

Subfamily AULODRILINAE

+*Limnodriloides winckelmanni* Michaelsen

+GENUS *Thalassodrilides* Brinkhurst and Baker

Subfamily RHYACODRILINAE

Monopylephorus rubroniveus Levinsen

Monopylephorus parvus Ditlevsen

Monopylephorus irroratus (Verrill)

+GENUS *Rhizodrilus* Smith

Subfamily PHALLODRILINAE

Phallodrilus prostatus (Knöllner)

Phallodrilus parthenopaeus Pierantoni

+*Phallodrilus profundus* Cook

Aktedrilus monospermathecus Knöllner

Spiridion insigne Knöllner

+GENUS *Bathydrilus* Cook

+GENUS *Peosidrilus* Baker and Erseus

+GENUS *Uniporodrilus* Erseus

Bacescuella arctica Erseus

+GENUS *Adelodrilus* Cook

OTHER GENERA: +*Smithsonidrilus* Brinkhurst
+*Coralliodrilus* Erseus
+*Heterodrilus* Pierantoni
+*Jolydrilus* Marcus

Family ENCHYTRAEIDAE (No generally accepted subfamilial classification)

+*Achaeta littoralis* Lasserre

Cernosvitoviella immota (Knöllner)

Cognettia glandulosa (Michaelsen)

Cognettia sphagnetorum (Vejdovsky)

Enchytraeus albidus Henle

Enchytraeus buchholzi Vejdovsky

Enchytraeus capitatus Bulow emm. Nielsen and Christensen

Enchytraeus minutus Nielsen and Christensen

+*Enchytraeus liefdeensis* Stephenson

Fridericia callosa (Eisen)

Fridericia paroniana Issel

Fridericia perrieri (Vejdovsky)

Fridericia ratzeli (Eisen)

Fridericia striata (Levinsen)

+*Grania macrochaeta pusilla* Erseus

Grania maricola Southern

+*Grania ovitheca* Erseus

+*Grania postclitellochaeta* (Knöllner)

+*Grania roscoffensis* Lasserre

+*Grania variochaeta* Erseus and Lasserre
+*Hemifridericia parva* Nielsen and Christensen
Henlea nasuta (Eisen)
Henlea perpusilla Friend
Henlea ventriculosa (Udekem)
Lumbricillus arenarius (Michaelsen)
Lumbricillus bulowi Nielsen and Christensen
Lumbricillus christenseni Tynen
Lumbricillus dubius (Stephenson)
+*Lumbricillus fennicus* Nurminen
Lumbricillus helgolandicus (Michaelsen)
Lumbricillus kaloensis Nielsen and Christensen
Lumbricillus knollneri Nielsen and Christensen
Lumbricillus lineatus (Müller)
Lumbricillus niger Southern
Lumbricillus pagenstecheri (Ratzel)
Lumbricillus pumilio Stephenson
Lumbricillus reynoldsoni Backlund
Lumbricillus rivalis Levinsen emm. Ditlevsen
Lumbricillus scoticus Elmhirst and Stephenson
Lumbricillus semifuscus (Claparede)
Lumbricillus tuba Stephenson
Lumbricillus viridis Stephenson
Marionina achaeta (Hagen)
Marionina appendiculata Nielsen and Christensen
Marionina arenaria Healy
Marionina argentea (Michaelsen)
Marionina communis Nielsen and Christensen
Marionina preclitellochaeta Nielsen and Christensen
Marionina sjaelandica Nielsen and Christensen
Marionina southerni (Cernosvitov)
Marionina spicula (Leuckart)
Marionina subterranea (Knöllner)
Mesenchytraeus armatus (Levinsen)

Phylum Annelida incertae sedis

Family AEOLOSOMATIDAE

+*Aeolosoma maritimum* Westerheide and Bunke
+*Aeolosoma litorale* Bunke

Systematic section

Key to families

1. Setae robust, usually few in each bundle (or absent), straight or sigmoid, simple-pointed. Cuticle thick. Spermathecae paired in V. (Testes in XI, ovaries and male pores in XII) .... ENCHYTRAEIDAE (p. 91)

Setae variable, all bifid or with hair setae dorsally or hair setae in all bundles. Cuticle thin. Spermathecae in testes segment, or several pairs of unicellular spermathecae in anterior of body **2**

2. Move via ciliated prostomium, septa absent. Hair setae in dorsal and ventral bundles, with simple-pointed, smooth or toothed sigmoid setae in some posterior bundles, often only ventrally, or absent, or all setae absent (free-living forms). Spermathecae unicellular, several pairs anteriorly. Ovaries single, mid-body; testes several pairs in front of and behind ovary. (Very small, less than 4 mm . . . AEOLOSOMATIDAE (p. 112)

Move by contractions of muscular segmented, septate body. Hair setae dorsally or absent. Spermathecae paired or single, multicellular, in testicular segment preceding ovarian segment (gonads paired) which contains male duct **3**

3. Asexual reproduction forming chains of zooids. 'Eyes' sometimes present. Usually less than 20 mm long. Male pores in ovarian segment (one segment between V and VIII), spermathecae in segment immediately in front of that (i.e. testicular segment). Pectinate setae rarely present NAIDIDAE (p. 27)

Asexual reproduction by simple fission when present. No eyes. Usually longer than 20 mm but offshore and meiobenthic marine species small. Male pores in ovarian segment, usually in XI, spermathecae in X. Pectinate setae often present dorsally, almost invariably accompanied when present by hair setae, *or* setae all bifid, *or* with hairs and simple-pointed setae dorsally, especially in posterior segments of marine species .... TUBIFICIDAE (p. 40)

Table 1. *Summary of families*

Family	Size	Eyes	Setae	Testes (segment)	Ovaries (segment)	Spermathecae (segment)	Male pore (segment)	Asexual reproduction	Habit
Naididae	Usually < 20 mm	Present or absent	All bifid *or* dorsals absent or simple-pointed *or* hairs and needles dorsally, ventrals bifid. Numerous in each bundle. Often missing from some anterior dorsal bundles	IV or V or VII	V or VI or VIII	IV or V or VII	V or VI or VIII	Normal mode of reproduction – forms chains (paratomy) of zooids	Burrow or swim
Tubificidae	Usually larger than 20 mm, marine species small	Absent	All bifid *or* dorsals accompanied by hair setae *or* dorsals with hair and pectinate setae from II, numerous in each bundle	X	XI	X	XI	Unusual, by fission when observed	Burrow
Enchytraeidae	Usually larger than 20 mm, marine species small	Absent	All simple-pointed, few in number (bifid in *Propappus*, absent in *Achaeta*, missing from many segments in *Grania*)	XI	XII	V	XII	Presumed absent	Burrow
Aeolosomatidae	Chains of up to 10 mm, usually much less; Individuals up to 2 mm	Absent	Absent, *or* hair setae in all bundles, sometimes with toothed or smooth sigmoid needles limited to posterior bundles, sometimes ventral only	In several segments, paired, both sides of ovary	Single, median mid-body	2–5 pairs, single cells, anterior half of body	(Use posterior nephridia)	Normal mode of reproduction – forms chains of zooids	Swim via cilia on prostomium, septa absent in most

All other families have two setae per bundle, all alike, with few exceptions, with the addition of the aquatic Phreodrilidae from the southern hemisphere (hair setae from III dorsally, simple or bifid ventrals, often one of each per bundle).

Family NAIDIDAE

The naidids are perhaps the most completely adapted aquatic oligochaetes, having rudimentary eyes and some capability for swimming. The setae are quite diverse in form, one common characteristic being their absence from the first five or six dorsal bundles in many genera. The ventral setae of the same bundles are often different to those of the posterior segments, these differences being produced as a result of the asexual reproduction process that dominates the life history. These delicate worms are usually smaller than those of the most closely related family, the tubificids, apart from the marine species of that family. The smallest marine tubificids frequently have bifid setae with reduced upper teeth, a form not found in the Naididae.

Most of the species included here are freshwater forms sometimes recorded in slightly brackish waters. Most of these records are from the Baltic, where many freshwater forms have become resistant to minimal salt levels. The genus *Paranais* (plus *Wapsa*, which is possibly synonymous) contains species found in truly marine coastal situations, but none of the family is found on the sea-bed well offshore, where tubificid species can be found. *Paranais* is undoubtedly present in Britain and *Amphichaeta sannio* has been recorded in the Forth (McLusky *et al.*, 1981). The other species listed have been found, along with other freshwater species, in some British estuaries where salt may penetrate the water column but not the sediments (Brinkhurst, 1963*c*).

Key to genera

1. No dorsal setae present .... *Chaetogaster* (p. 28)

Dorsal setae present **2**

2. Hair setae absent from all setal bundles **3**

Hair setae present in dorsal setal bundles **5**

3. Dorsal setae from III onwards .... *Amphichaeta* (p. 30)

Dorsal setae from V or VI **4**

4. Dorsal setae from VI .... *Uncinais* (p. 34)

Dorsal setae from V .... *Paranais* (p. 32)

5. Anterior end forming a proboscis .... *Stylaria* (p. 38)

No proboscis on the prostomium .... *Nais* (p. 36)

Genus CHAETOGASTER von Baer

(Fig. 7)

This genus is immediately recognizable by the absence of dorsal setae. The ventral setae are bifid (except in *Chaetogaster setosus* Svetlov). The nomenclature is complex with many synonyms, and species identification may be troublesome as the characters used require considerable attention to detail. This is really a freshwater genus, with *Chaetogaster limnaei* von Baer being commensal with, or parasitic upon, a variety of molluscs (sometimes in brackish water).

The species reported from brackish water are:

Chaetogaster langi Bretscher

(Fig. 7*a*,*d*)

0.9–2 mm long. Prostomium inconspicuous, without median incision. Setae of II 3–9 per bundle, 3–6 in the rest; those of II less than 100 μm long.

Chaetogaster diaphanus Gruithuisen

(Fig. 7*b*)

2.5–25 mm long. Prostomium inconspicuous, without median incision. Setae of II 6–12 per bundle, 4–10 in the rest; those of II greater than 145 μm long.

A third species, *Chaetogaster crystallinus* Vejdovsky (Fig. 7*c*), has a median incision in the prostomium, and a few other characteristics which are said to differ from those in *C. diaphanus*. Kasprazak (1972*a*), however, has claimed that the two are synonymous.

All three have been recorded from Britain.

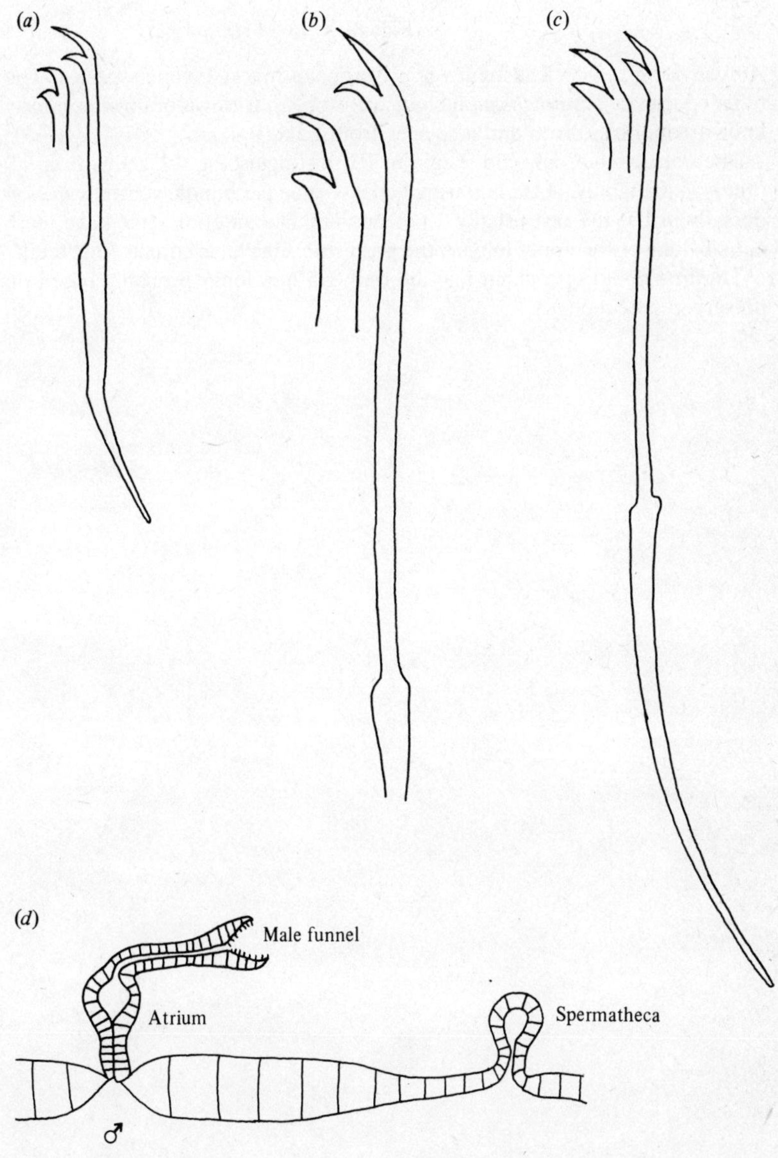

Fig. 7. Setae and reproductive organs of *Chaetogaster*: setae from the ventral bundles of II (left-hand example) and VI (right-hand example) of (*a*) *C. langi*, (*b*) *C. diaphanus* and (*c*) *C. cristallinus*; and (*d*) reproductive organs of *C. langi*. (After Sperber, 1948.)

Genus AMPHICHAETA Tauber

(Fig. 8)

Amphichaeta sannio Kallstenius is a European brackish water species. The other species of *Amphichaeta* include a European freshwater form, a poorly known American taxon and a species from Lake Baikal.

The dorsal setae, all bifid, begin in III, distinguishing the genus from all others in the family. *A. sannio* usually has 4 setae per bundle ventrally in II, 4 dorsally in III, the rest usually 3 per bundle. The anterior setae have teeth equally long or the upper longer; the posterior setae have equally long teeth.

The first zooid of a chain may be only 1.5 mm long (probably based on preserved specimens).

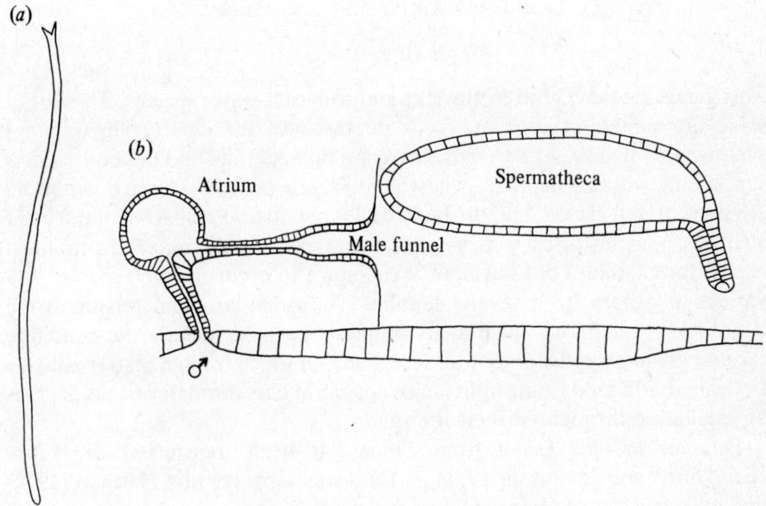

Fig. 8. *Amphichaeta sannio:* (*a*) seta and (*b*) reproductive organs. (After Sperber, 1948.)

Genus PARANAIS Czerniavsky

(Fig. 9)

This genus contains both freshwater and truly salt-water species. The dorsal setae are bifid and start in V, as do those of the closely allied if not synonymous *Wapsa*. In *Paranais*, there are no nephridia and the body wall is supposedly without papillae; whereas in *Wapsa evelinae* Marcus, nephridia are present, but closed, and the body wall is papillate. Papillation of the body wall was used historically as a generic character (in *Peloscolex*, a tubificid genus, for example) but has now been shown to occur in some species in a variety of genera from several families. A further problem relates to the degree of papillation, which at its simplest seems to include the condition created presumably by body wall secretions, in which foreign matter adheres to the body. Indeed, some individuals appear to pass through various degrees of papillation throughout their life span.

Paranais mobilis Liang from China has been transferred to *Wapsa* (Brinkhurst and Jamieson, 1971) and this was supported by Harman (1977) in his description of *W. grandis* from the USA. The species is supposed to have encrusted foreign matter on the body wall rather than papillae. The same feature is present, however, in *Paranais frici* Hrabe, which the above authors have left in *Paranais*. W. J. Harman (personal communication) suggests that a subgeneric separation might be more reasonable but argues for retaining the status quo as much as possible. As the name *Paranais* has priority, the names used here should remain stable unless both genera are retained and *P. frici* proves to be a *Wapsa* species.

The distinction between the species is as follow:

Paranais litoralis (Müller)

(Fig. 9*a*,*c*)

2–3.5 mm preserved length. Dorsal setae slightly thinner than ventrals, ventrals of II 5–7 per bundle, slightly longer than the rest, upper tooth longer than the lower. In the rest, usually 2–3 per bundle, upper tooth as long as or only slightly longer than the lower. Body wall naked. Known from Kent and Norfolk.

Paranais frici Hrabe

(Fig. 9*b*)

2.7–5.7 mm preserved length. Ventral setae of II 2–4 per bundle, upper tooth more than twice as long as the lower. In other segments 1–2 per bundle (3 dorsally in V) with the upper tooth longer than the lower. Body wall encrusted with foreign matter. Not yet recorded in Britain.

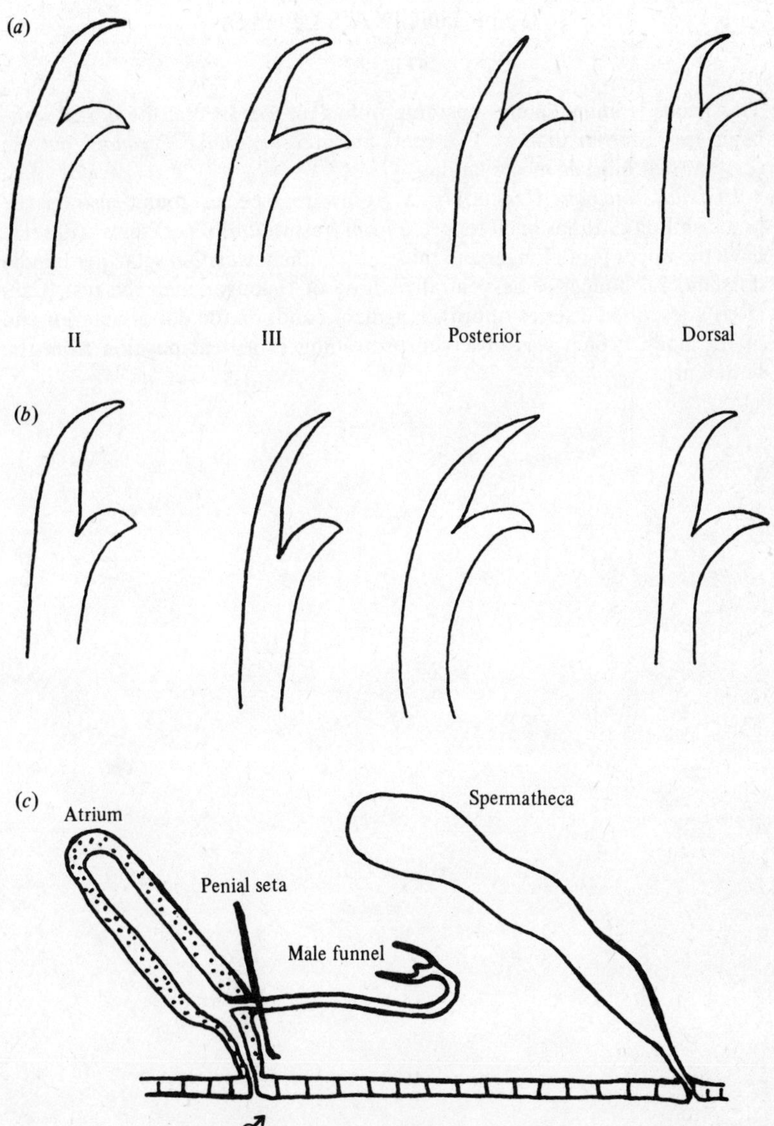

Fig. 9. *Paranais* spp.: ventral seta of II, III, and a posterior bundle, and a dorsal seta of (*a*) *P. litoralis* and (*b*) *P. frici* (after Sperber, 1948); (*c*) reproductive system of *P. litoralis* (after Knöllner, 1935).

Genus UNCINAIS Levinsen

(Fig. 10)

This genus is immediately separable from *Paranais* in that the dorsal setae begin in VI rather than V. Eye spots are present, unlike *Paranais*, but the setae are all bifid as in the latter.

Uncinais uncinata (Orsted) is a freshwater species found also in the brackish Baltic. It has been reported from freshwater in Yorkshire. All setae have the upper tooth longer and thinner than the lower; 2–4 setae per bundle dorsally, 2–7 longer setae ventrally, those of II longer than the rest. Cain (1959) described a series of brown pigment bands on the dorsal anterior end of the body, which was observed protruding in a bent position from the sediment.

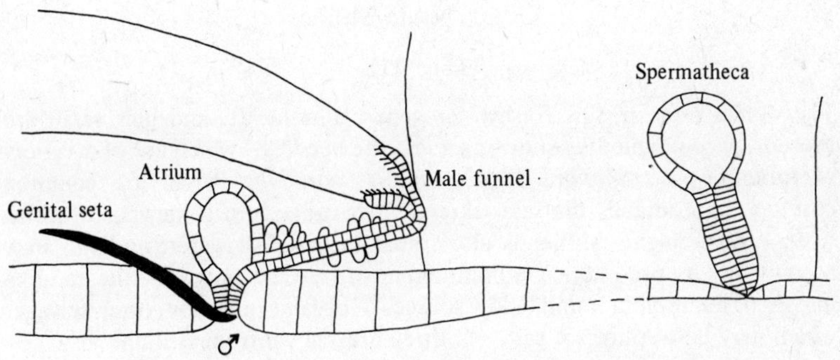

Fig. 10. Reproductive system of *Uncinais uncinata* (partly after Sperber, 1948).

Genus NAIS Müller

(Fig. 11)

The dorsal setae in this freshwater genus start in VI, and hair setae are present, accompanied by shorter setae – the needles – which are of a variety of forms. *Nais communis* Piguet and *N. variabilis* Piguet are common cosmopolitan animals that can extend their range into somewhat brackish water. *Nais elinguis* Müller is also found in brackish water, and this may account for its presence as a component of 'sewage fungus' – the rag-like masses of microbiota found below sources of organic pollution, often sewage, which may be a source of salt. All three are known from Britain.

The first two species are very variable in size, size of setae, pigmentation, and setal shape (Sperber, 1948) and these species are hard to separate. Long lists of synonyms exist, and biologists should not be surprised to find the separation difficult. *Nais elinguis* (Fig. 11*e,f*) is readily distinguishable by its needles, which have two long, parallel teeth clearly visible in one orientation, but not at others when the needles will appear to be simple-pointed or even to have two short teeth.

In *N. variabilis* and *N. communis* the needle teeth are short but diverge very little. In the former the eyes are sometimes absent, the anterior ventral setae have the upper teeth about twice as long as the lower (Fig. 11*c,d*), in *N. communis* the ventral setal teeth are of equal length (Fig. 11*a,b*) and the eyes are supposed to be present. The so-called stomach of *N. communis* is said to widen gradually, rather than abruptly as in *N. variabilis*. The latter has the ability to swim, and has longer, thicker, hair setae than *N. communis*. Dr R. Grimm (personal communication) tells me that *N. communis* has longer setae than *N. variabilis*, the divergent needle teeth being clearly visible at × 40, whereas those of *N. variabilis* require the use of an oil immersion lens.

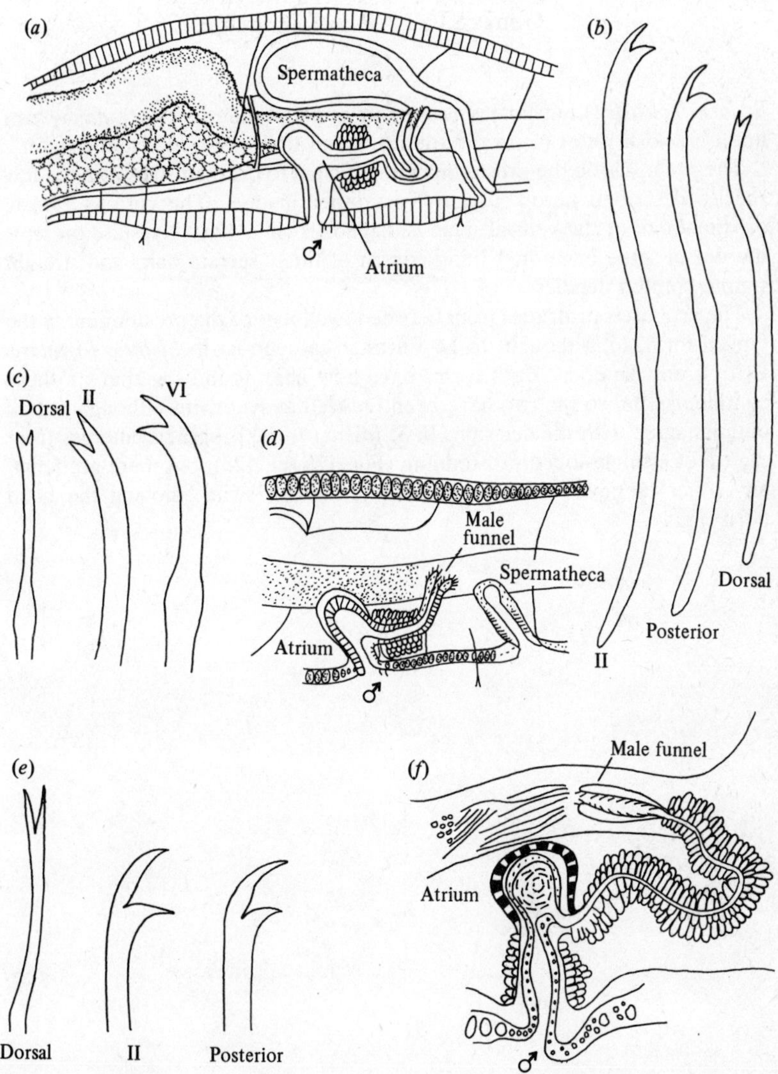

Fig. 11. *Nais* spp.: reproductive system and setae of *N. communis* (*a*, *b*), *N. variabilis* (*c*, *d*) and *N. elinguis* (*e*, *f*) (after Sperber, 1948). (The needles of *N. communis* are larger than those of *N. variabilis* – teeth visible at × 40 rather than × 100 – and the teeth are more divergent (R. Grimm, personal communication).)

Genus STYLARIA Lamarck

(Fig. 12)

Stylaria lacustris (Linnaeus) is a common, cosmopolitan species that may turn up in brackish water (as in the Baltic). It has been recorded from Britain.

The proboscis on the prostomium is sufficient to distinguish it from the few species described here, but not from other naidids. The curious zig-zag proximal end of the ventral setae is diagnostic (Fig. 12*a*). Eyes are present, the dorsal setae begin in VI and consist of finely serrate hairs and straight simple-pointed needles.

The proboscis protrudes from between two lobes of the prostomium in the typical form, long thought to be Eurasian in contrast to *Stylaria fossularis* Leidy from America. Both forms have now been found together on these continents, and so the two have been reduced to synonyms although not all authors agree with the decision. In *S. fossularis* the proboscis emerges from the tip of a single-lobed prostomium (Fig. 12*b*, cf. 12*c*), but there are few if any other tangible differences. Both forms can swim, moving the head horizontally.

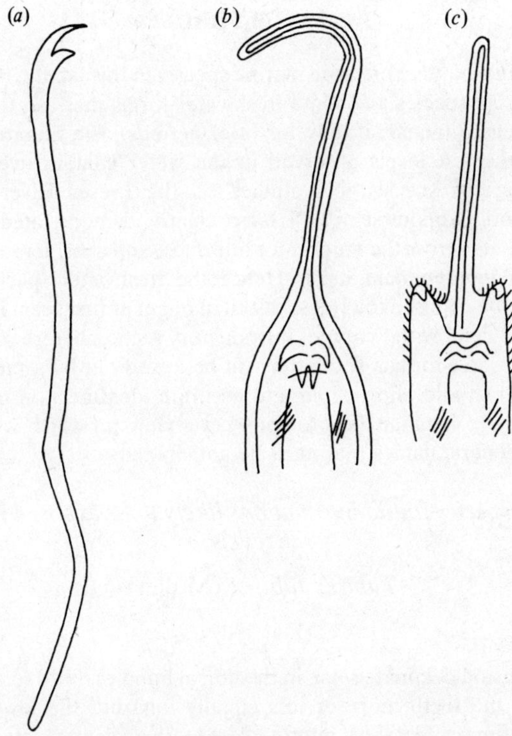

Fig. 12. *Stylaria lacustris*: (*a*) ventral seta; (*b*, *c*) the form of the proboscis in '*S. fossularis*' (*b*) and in '*S. lacustris*' (*c*). (Partly after Sperber, 1948.)

Family TUBIFICIDAE

There are many genuine offshore marine species in this family, in addition to coastal/estuarine species and some freshwater forms that can tolerate some salt in their environment, if only for brief periods. The literature is full of citations of the salt levels observed in the water column over sediments containing various species, but studies on the Fraser River in Canada (Chapman and Brinkhurst, 1981) have clearly demonstrated very steep salinity gradients across the mud/water interface, and considerable constancy of salinity in the sediment itself. Hence the freshwater species found in estuaries may be exposed to far less salt than might at first seem likely. These species will be dealt with in a brief introductory section before proceeding to the truly salt-water forms. The latter will be keyed out by a rather unusual method which should allow rapid and accurate identification more readily than by the more familiar dichotomous key style, and which will withstand the shifts in generic names that must be anticipated.

Freshwater species found in situations likely to be exposed to very low salt levels

Tubifex tubifex (Müller)

(Fig. 13*a*)

Has hair setae and pectinate setae in the dorsal bundles, bifid setae in ventral bundles with the teeth more or less equally long but the lower distinctly thicker, sometimes a few short intermediate teeth on ventral setae. The penis sheaths are short, tub-shaped, thin and rugose. There are no genital setae (on mature specimens). The hair setae may be serrate. Unusual forms with simple hair and bifid setae dorsally, or all bifid setae, may be encountered. The species is common in polluted water or marginal habitats not occupied by other species. It is not common in associations of several species, though it is cosmopolitan and widely distributed in Britain.

Limnodrilus hoffmeisteri Claparede

(Fig. 13*b*)

Is the commonest aquatic oligochaete in most localities, and is found further seaward than any other truly freshwater species. The setae are all bifid, with the teeth of variable proportions. The penis sheaths in XI are elongate cylinders of thickened cuticle up to 14 times longer than broad, with a broadened head (distal end), often with the opening at right angles to the shaft. The penes are surrounded by spirally arranged muscles. There are no genital setae in mature forms. *L. hoffmeisteri* is widely distributed in Britain, and its ecology in the Thames (together with that of *Tubifex tubifex*) is documented by Birtwell and Arthur (1980).

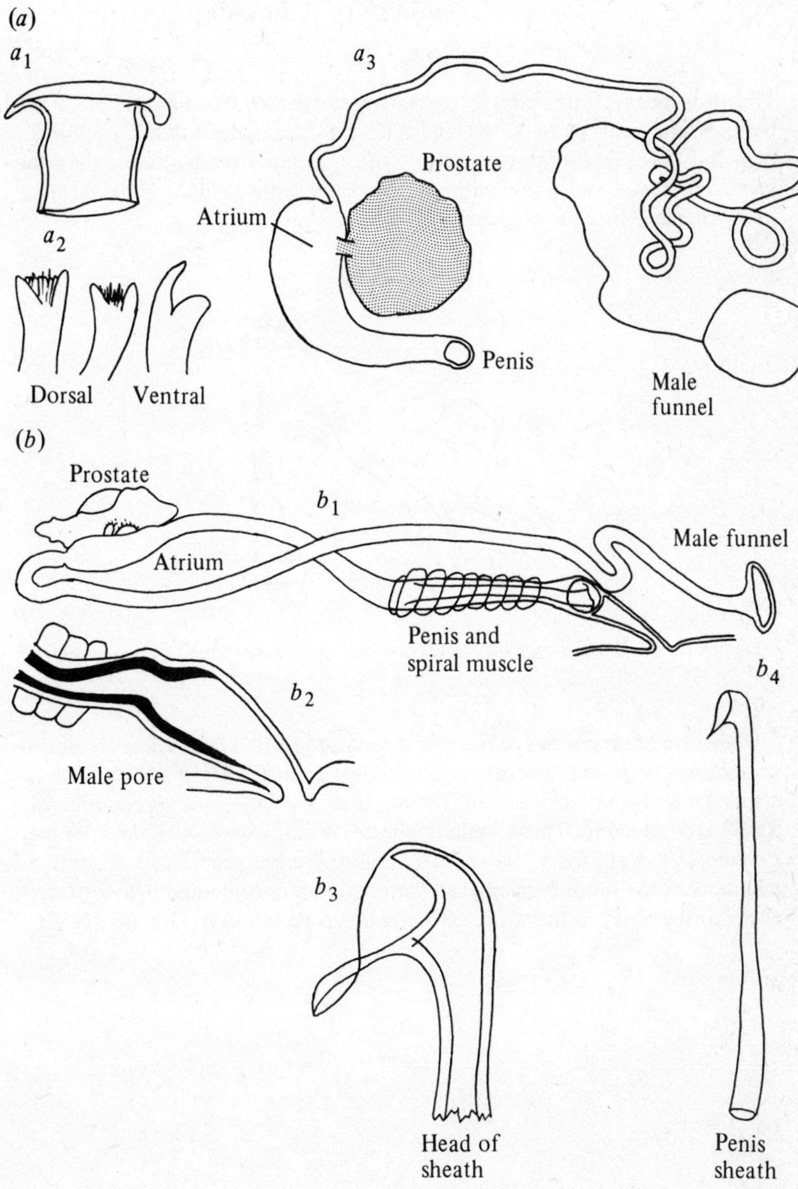

Fig. 13. Essentially freshwater tubificids capable of limited penetration into brackish and estuarine waters. (*a*) *Tubifex tubifex* (a_1, penis sheath; a_2 setae; a_3, reproductive system); (*b*) *Limnodrilus hoffmeisteri* (b_1, b_2, reproductive system; b_3, b_4, penis sheath). (After Brinkhurst and Jamieson, 1971, and Stephenson, 1912, respectively.)

Isochaetides michaelseni (Lastockin)

(Fig. 13*c*)

This European, fresh- and brackish-water species has bifid setae in all bundles, the ventrals of X, XI and XIII with the upper tooth much longer than the lower, those more anterior with the upper teeth shorter than the lower. There are long penes without cuticular sheaths, and the male ducts are quite characteristic of the genus (Fig. 13*c*).

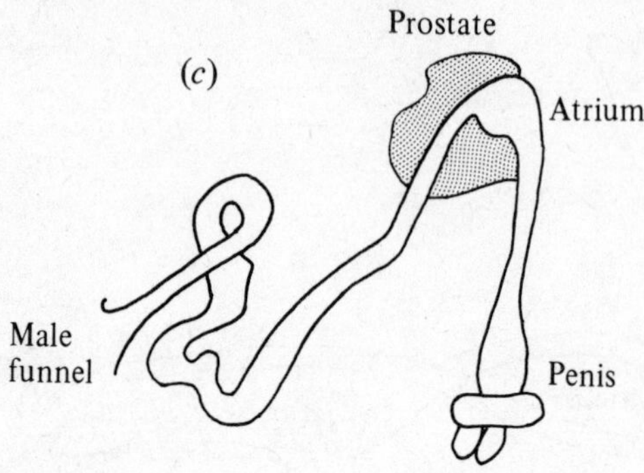

Fig. 13 (*c*). *Isochaetides michaelseni*, reproductive system (after Lastockin, 1937).

There are other species in the genus reported from inland saline localities. In addition there are several species known from the Caspian Sea and the estuaries of the Black Sea in genera that are otherwise predominantly freshwater in habit. These include species within *Psammoryctides* Hrabe, *Potamothrix* Vejdovsky and Mrazek, and *Tubifex* Lamarck, as well as members of the more typically salt-water genera documented below. References to the Russian literature concerned can be found in Grigelis (1980).

Guide to the identification of salt-water Tubificidae

Decision Level 1

A. Dorsal anterior setae from II–IV pectinate, those of V to about XIV broadly palmate, all other setae bifid (Fig. 14*b*). Cuticular penis sheaths tub-shaped. No hair setae, no genital setae. Europe, including the Black Sea .... *Tubifex costatus* (Claparede) (p. 52)

B. In II, III and IV, two straight setae with proximal hooks in each bundle, increasing in size posteriorly. From V on, one large, slightly hooked seta in each 'bundle' (i.e. four per segment). Penial setae in ventral bundles of XI (mature specimens) single, straight, simple-pointed. West Norway *Bathydrilus rarisetis* (Erseus) (p. 80)

C. Two trifid or blunt setae in each anterior bundle. Posterior setae single, upper tooth thin and shorter than the lower. Ventral setae of XI paired rod-like penials. Mediterranean *Heterodrilus arenicolus* Pierantoni (p. 88)

(Other species with trifid setae found in North America and Australia originally assigned to this species or the subspecies *queenslandicus* may now prove to be distinct, possibly even at the generic level)

D. Dorsal setae hairs and pectinate setae from II .... **Level 2(i)** (p. 44)

E. Hair setae from II or beginning as far back as the middle segments. Other dorsal setae bifid from II, sometimes simple-pointed from VII or IX or more posteriorly. (One species with occasional faintly pectinate setae) **Level (ii)** (p. 45)

F. All setae bifid (sometimes simple-pointed posteriorly) or modified genital setae in some or all ventral bundles of IX–XIII in mature forms **Level 2(iii)** (p. 46)

Level 2(i)

Species with hair and pectinate setae from II.

A. Body wall naked. Hair setae with lateral hairs. Dorsal pectinate setae with thin U-shaped teeth and a few, fine intermediate teeth. Upper teeth of ventral setae twice as long as the lower. Penis sheaths short, tub-shaped. Atlantic boreal .... *Tubifex nerthus* Michaelsen (p. 52)

B. Body wall naked, with secretory material to which foreign matter adheres. Hair setae with short, sparse lateral hairs. Pectinate setae with long parallel teeth and 1–2 thin intermediate teeth. Upper teeth of ventral setae exceedingly thin, both teeth elongate. Penis sheath with lateral protuberance. Abyssal, Northwest Atlantic. (39°46.5′N, 70°43.3′W)
Tubificoides aculeatus (Cook) (p. 56)

C. Body wall naked. Pectinate setae robust, with short stout teeth; hair setae often serrate. Penis sheaths thin, tub-shaped, rugose
Tubifex tubifex (Müller) (p. 40)

Note: *Monopylephorus irroratus* has a few, twisted hair setae and bifid setae dorsally, but pectinate setae are encountered rarely. See Level 2(ii).

Other species: *Variechaeta israelis* Brinkhurst – salt springs, Israel

Tubificoides (?) debilis (Finogenova) – Caspian Sea

Tubificoides nerthoides (Brinkhurst) – North America, coastal Atlantic and Pacific

Tubificoides (?) maritimus (Hrabe) – Black Sea; anterior setae not described

Level 2(ii)

Species with hair setae and bifid setae dorsally.

A. Body wall naked. Hair setae from II thin, twisted distally, 1–2 per bundle. Up to 4 bifids dorsally, up to 7 ventrally, upper teeth longer than the lower, rarely an intermediate tooth dorsally. Coelomocytes abundant. Penes eversible, cuticular process on lining. Atlantic, coastal-estuarine *Monopylephorus irroratus* (Verrill) (p. 66)

B. Body wall finely granulate behind clitellum. Dorsal setae bifid from II to VI or VII with teeth sub-equal, 2–4 bent hair setae without lateral hairs. From VII or VIII posteriorly 1–3 hairs and 1–3 hair-like setae. Ventral setae 2–3 anteriorly, with upper tooth slightly the larger, posteriorly 1–2. Penis sheaths thin truncate cones. No coelomocytes. West Norway; Skagerrak, Sweden. . . . . *Tubificoides amplivasatus* (Erseus) (p. 56)

Note: Rare specimens of *Tubificoides benedeni* (Udekem) have a few hair setae according to some accounts. These may belong to a separate species (see p. 55).

Other species: *Jolydrilus jaulus* Marcus – Brazil

Tubificoides (?) maritimus (Hrabe) – Black Sea

Tubificoides euxinicus (Hrabe) – Black Sea

Tubificoides swirencowi (Jaroschenko) – Black Sea

Tubificoides intermedius (Cook) – Massachusetts, USA

Tubificoides apectinatus (Brinkhurst) – USA, Canada, Pacific and Atlantic

Tubificoides dukei (Cook) – North Carolina, USA

Tubificoides postcapillatus (Cook) – Mexico

Level 2(iii)

Species without hair setae dorsally.

A. Gut in immediate pre-clitellar region, usually in segment IX, with two blind diverticula attached to the intestine posteriorly and extending forward to 8/9. Spermathecal setae present (ventral bundles of X) in one European species. .... *Limnodriloides* Pierantoni (p. 59)
Smithsonidrilus Brinkhurst (p. 88)

B. No gut diverticula. Modified genital setae replace normal ventral setae in mature specimens from some or all of IX–XIII. No species in Europe with cuticular penis sheaths. None with papillate body wall. .. **Level 3** (p. 47)

C. No gut diverticula. No genital setae on mature specimens. Some species with cuticular penis sheaths. Some species with papillate body wall.
Level 4 (p. 48)

Level 3

Species with modified genital setae but without papillae or cuticular penis sheaths. No hair setae.

Genital setae replace normal ventral setae in mature specimens on segments:

A. X and XI (spermathecals paired with upper teeth 2.5 times longer than the lower, penials paired, straight, simple-pointed). Gulf of Naples, Scotland .... *Phallodrilus parthenopaeus* Pierantoni (p. 74)

B. XI only, in two sizes *Adelodrilus* Cook (p. 86); *Inanidrilus* Erseus (p. 71)

C. XI only, 1–22 similar setae .... *Phallodrilus* Pierantoni (p. 72); *Heterodrilus* (p. 88); *Spiridion* Knöllner (p. 78)

D. X and XII (simple-pointed setae singly according to one description). Cosmopolitan, coastal. .... *Clitellio arenarius* Müller (p. 58)

E. IX, X and XI, two long thick single-pointed setae on each side ventrally. Anterior somatic setae with rudimentary upper teeth, the rest simple-pointed. Atlantic abyssal (38°08′–46′N, 68°–71°W)

Phallodrilus profundus Cook (p. 72)

F. X, XI, XIII (none in XII) bifid with upper tooth much longer than lower. Freshwater and brackish water. Europe

Isochaetides michaelseni (Lastockin) (p. 42)

Note: Two *Limnodriloides* species have genital setae and will appear here if the gut diverticula are not observed.

Other species: IX and XI: *Rhizodrilus lowryi* (Cook) – Antarctic; *Bathydrilus rohdei* (Jamieson) – Great Barrier Reef; *Rhizodrilus pacificus* Brinkhurst and Baker – Canada Pacific

X only: *Isochaetides hamata* (Moore) – Massachusetts, USA; *Isochaetides suspectus* (Sokolskaja) – Far Eastern USSR

XI only: 1–22 similar: *Monopylephorus frigidus* Brinkhurst – Alaska, USA; *Bathydrilus adriaticus* (Hrabe) – Adriatic; *Peosidrilus* Baker and Erseus (p. 82); *Coralliodrilus* (p. 90); *Bermudrilus* (p. 86); *Uniporodrilus* (p. 82)

Level 4

No hair setae. No gut diverticula. No genital setae.

A. Mature specimens with a pair of cuticular penis sheaths in XI, sometimes thin and indistinct .... **Level 4(i)** (p. 49)

B. Mature specimens with penis in XI without specially thickened cuticular sheaths .... **Level 4(ii)** (p. 50)

C. Mature specimens without true penes .... **Level 4(iii)** (p. 51)

Level 4(i)

No hair setae, no gut diverticula, no genital setae, with cuticular penis sheaths.

A. Worms evenly and closely papillate. Somatic setae only two per bundle, with rudimentary upper tooth or simple-pointed. Penis sheaths cylindrical with recurved distal end. Boreal, coastal-estuarine *Tubificoides benedeni* (Udekem) (p. 55)

B. Non-papillate; 3–4 or 6 bifid setae with upper teeth thinner than and as long as (or shorter than) the lower teeth. Penis sheaths somewhat conical. Boreal, coastal-estuarine ..... *Tubificoides pseudogaster* (Dahl) (p. 55)

C. Non-papillate; 2–4 setae per bundle, bifid with teeth about equally long and thick. Penis sheaths very thin, more or less short cylinders. North Iceland .... *Tubifex litoralis* Erseus (p. 52)

D. Papillate to non-papillate. Anterior setae 5 per bundle, bifid. 1–2 simple-pointed setae from XII. Boreal, coastal, estuarine *Tubificoides heterochaetus* (Michaelsen) (p. 55)

E. Non-papillate. Setae usually about 7 per bundle anteriorly, bifid. Penis sheaths long, thick, very distinctive. Cosmopolitan, freshwater to upper estuarine .... *Limnodrilus hoffmeisteri* Claparede (p. 40)

Other species: *Tubificoides longipenis* (Brinkhurst) – Atlantic Boreal America

Tubificoides gabriellae (Marcus) – North and South America, Atlantic and Pacific shores

Tubificoides diazi Brinkhurst and Baker – New Jersey, USA

Tubificoides wasselli Brinkhurst and Baker – Delaware, USA

Tubificoides brownae Brinkhurst and Baker – Delaware, USA

Tubificoides maureri Brinkhurst and Baker – Delaware, USA

Tubificoides coatesae Brinkhurst and Baker – British Columbia

Tubifex (?) acapillatus Finogenova – Caspian Sea

Level 4(ii)

No hair setae. No gut diverticula. No genital setae. No cuticular penis sheaths, but true penes.

A. 2–3 setae per bundle, more or less bluntly simple-pointed or with rudimentary upper teeth (possibly single, thick, blunt ventral setae on X and XII). Spermathecae present, with spermatozeugmata. No prostate gland. Boreal, coastal . . . . *Clitellio arenarius* (Müller) (p. 58)

B. Three bifid setae anteriorly, upper tooth about as long as lower, 2–3 posteriorly with upper tooth shorter and thinner than lower. Spermathecae open near 9/10, ampullae, one in sperm sac in IX, one in X; sperm in bundles. Abyssal Northwest Atlantic

Bathydrilus asymmetricus Cook (p. 80)

C. 3–4 (rarely 5) setae per bundle, then 2–3 falling to one posteriorly, upper teeth shorter and thinner than the lower. Single dorsal spermatheca, sperm in random masses. Atria with two prostates. Probably all Europe, Bermuda, ?East USA . *Aktedrilus monospermathecus* Knöllner (p. 76)

D. 2–6 setae per bundle, anteriorly and posteriorly, upper tooth shorter and basally thinner than lower. No spermathecae, external spermatophores attached to body wall in or near clitellar region by small stalk. Atria with two prostate glands. Iceland, North Norway, Scotland

Bacescuella arctica Erseus (p. 84)

Note: *Tubificoides heterochaetus* (Michaelsen) would have appeared here from literature accounts, but type material has cuticular penis sheaths.

Level 4(iii)

No hair setae. No gut diverticula. No genital setae. No true penes.

A. Anterior setal bundles with 3–4 or 5 setae, with upper tooth longer than or as long as the lower; posterior setae fewer, simple-pointed or with reduced upper teeth. Single left spermatheca with median ventral pore. Atria paired but single median male pore. Coelomocytes abundant. Cosmopolitan, coastal ...... *Monopylephorus parvus* Ditlevsen (p. 66)

B. Anterior bundles with 4–6 setae, upper tooth thinner than but as long as the lower; 2 setae per bundle posteriorly. Spermathecal and male pores paired within median invaginations at full maturity. Coelomocytes abundant. Cosmopolitan, coastal
Monopylephorus rubroniveus Levinsen (p. 66)

Other species: *Thalassodrilides belli* (Cook) – Mexico

Thalassodrilides gurwitschi (Hrabe) – Adriatic, Black Sea

Thalassodrilides milleri Brinkhurst and Baker – Delaware, USA

Spiridion modricensis (Hrabe) – salt spring, Yugoslavia, beside the Adriatic

Jamiesoniella athecata Erseus – Heron Island, Australia

Jamiesoniella bahamensis Erseus – Andros Island, Bahamas

Kaketio ineri Righi and Kanner – Bonaire, Antilles, Bermuda, Florida

Subfamily TUBIFICINAE

Coelomocytes absent, or at least not large and abundant. Single, stalked prostate glands on each atrium, rarely absent. Genital setae usually restricted to single, rarely paired, hollow-ended spermathecal seta in each ventral bundle of X, rarely penial setae of similar form, often missing. Spermatozeugmata consist of spirally arranged sperm with the heads along the core. True penes usually present, often with thickened cuticular sheaths.

The genus *Clitellio* was recently reinstated in this subfamily, and the genus *Tubificoides* has been re-erected and contains elements formerly found in *Tubifex* and *Peloscolex*. These two are fully marine, and there are coastal *Tubifex* species but few truly marine species in other genera.

Genus TUBIFEX Lamarck

(Figs. 14, 15)

This genus contains species which usually have quite elongate vasa deferentia, sometimes of two sections of differing width, which enter comma-shaped atria more or less apically. There are no distinct ejaculatory ducts, but the penes bear cuticular penis sheaths, usually of an abbreviated cylindrical form, sometimes conical. *Tubifex costatus* (Claparede) (Fig. 14*a*,*b*) has characteristic palmate setae in the dorsal bundles of about V–XIV, and is found in brackish waters all over Europe including the Ponto–Caspian river estuaries. *T. nerthus* Michaelsen (Fig. 14*c–f*) has hair setae and pectinate setae dorsally, the anterior ventral setae having quite long upper teeth. Both have short penis sheaths. The biology of *T. costatus* in the Thames estuary is detailed by Birtwell and Arthur (1980).

In *T. litoralis* Erseus (Erseus, 1976*a*) the setae are all bifid and the penis sheaths are very thin and poorly developed (Fig. 15*a–c*). The species is recorded from northern Iceland. It resembles *T. tubifex blanchardi* Vejdovsky, which was referred to as a euryhaline salt-water species by Grigelis (in Brinkhurst and Cook, 1980).

Some other former *Tubifex* are now listed in *Tubificoides*.

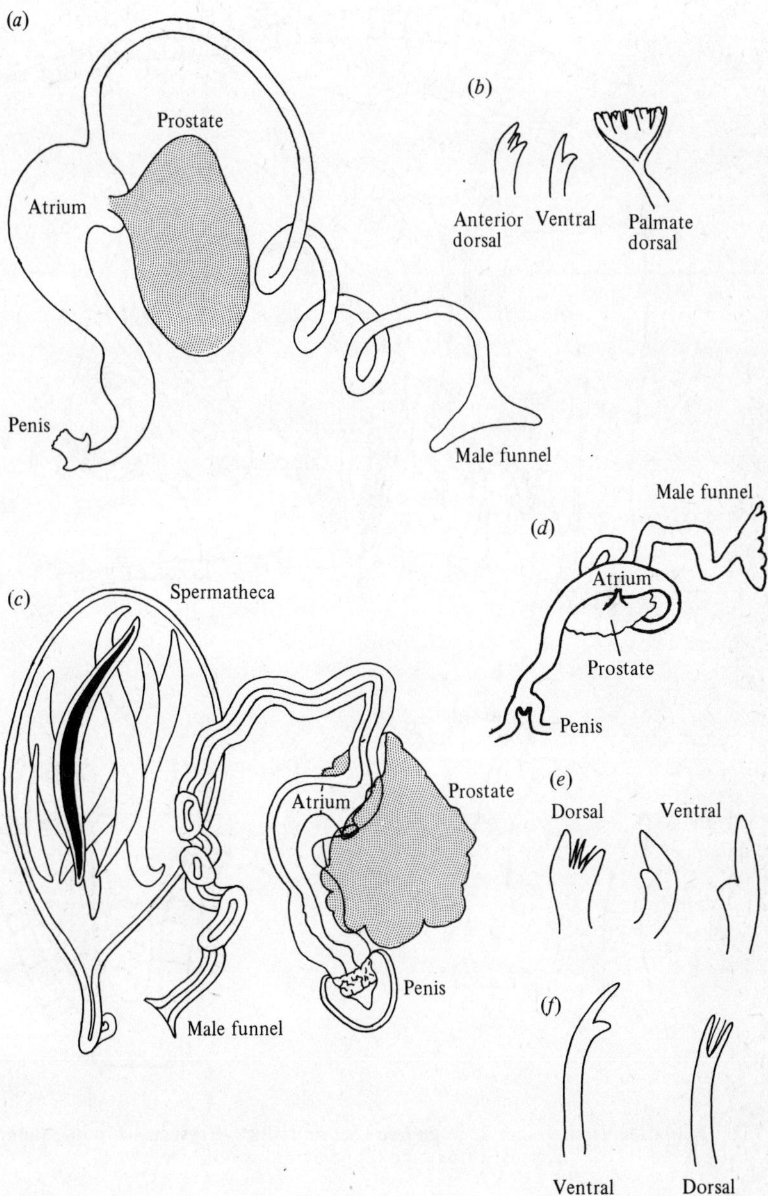

Fig. 14. Marine *Tubifex* spp. 1: reproductive system (*a*) and setae (*b*) of *T. costatus* (after Brinkhurst and Jamieson, 1971); *T. nerthus*, (*c*, *d*) reproductive system (*c* after Pickavance and Cook (1971) as '*T. newfyi*', *d* after Michaelsen, 1908) and (*e*, *f*) setae (after Brinkhurst and Jamieson, 1971).

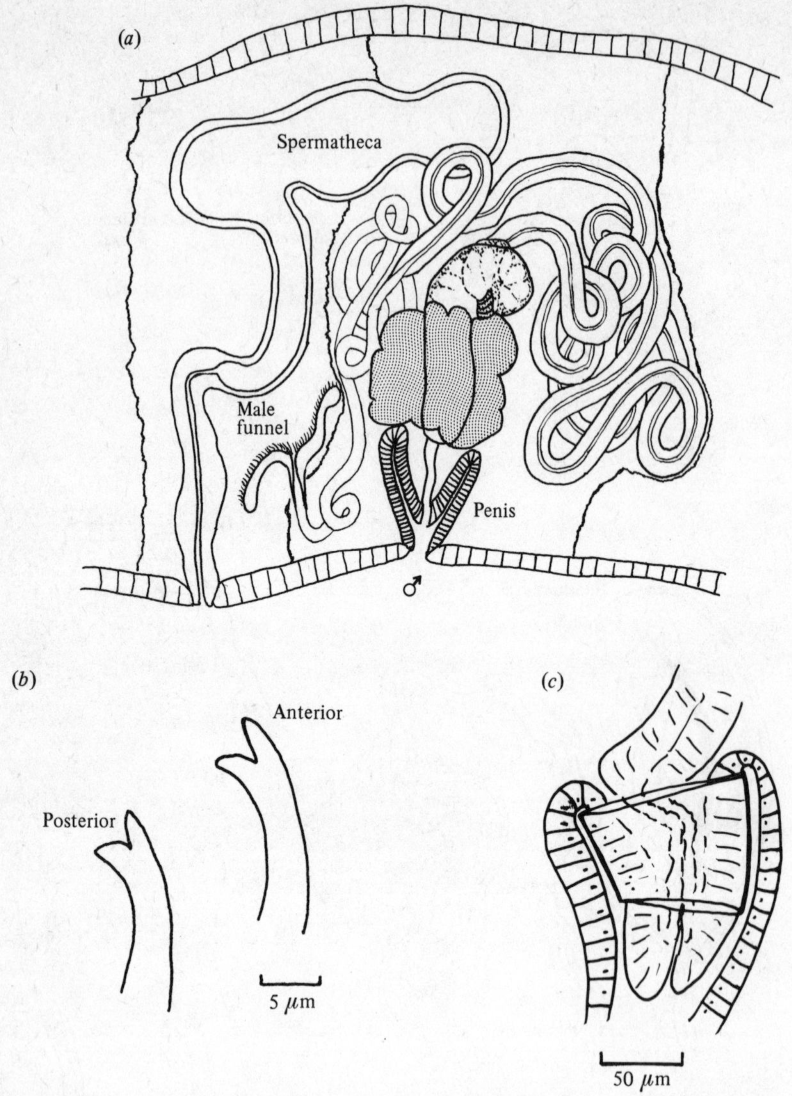

Fig. 15. Marine *Tubifex* spp. 2, *T. litoralis*: (*a*) reproductive system, (*b*) setae (after Erseus, 1976*a*) and (*c*) penis and sheath.

Genus TUBIFICOIDES Lastockin emm. Brinkhurst and Baker

(Fig. 16)

In a review of the marine Tubificidae of North America, Brinkhurst and Baker (1979) expanded on the re-erection of this genus by Holmquist (1978). There are now large numbers of species in the genus, many quite limited in distribution. Concern has been expressed about the possible cryptic speciation in species such as *Tubificoides gabriellae* and *T. pseudogaster*, which have not usually been well described at each locality from which they have been reported.

All tubificoids share quite characteristic, short, cylindrical atria with the vasa deferentia and prostates attached opposite each other subapically, and with the cap of the atria having a histological makeup distinct from the rest. The penes are lightly cuticularized and are of characteristic shape. In some, at least, that part of the atrium closest to the penis also differs from the rest.

Some of the species are papillate, which is a reference to both (*a*) the covering of the body wall with cuticular folds and to the secretions which usually involve foreign particles, and (*b*) the extensions of the body wall through this layer which can be presumed to provide the necessary tactile sense that would otherwise be blocked. These are sometimes referred to as secretory papillae (but this may simply be a translation error). This covering is well developed in *T. benedeni* (Udekem), a common coastal form in Britain (Fig. 1) often found with *Clitellio arenarius*. The former's biology has been discussed by Hunter and Arthur (1978) and Wharfe (1977) (under the name *Peloscolex*) working in the Thames-Medway estuary. *T. benedeni* has two setae in each anterior bundle, either bearing rudimentary upper teeth, or with these missing, presumably worn off. There are cuticular penis sheaths, though these cannot be seen in whole mounts (Fig. 16*a*,*b*). Holmquist (1978) considers this as the sole species in the genus *Edukemius* (which I regard as perhaps a subgenus) and prefers the original spelling *benedii* which has not been used in this century.

Tubificoides pseudogaster (Dahl) has more setae per bundle (up to six) but they are clearly bifid. The variations of the characters within what is currently regarded as a single species are documented in Brinkhurst and Baker (1979). The cuticular penes are cylindrical to thimble-shaped (Fig. 16*c*). A new American species differs in that some of the segments are particularly elongate and bear special annulations at the setal lines. Otherwise the species lacks any special coating on the body wall. A series of forms has been discovered in North America which may be separate species or perhaps local varieties of one taxon.

Tubificoides heterochaetus (Michaelsen) (Fig. 16*d*,*e*) has a few sparse papillae in some specimens. The dorsal setae are bifid anteriorly, simple-pointed posteriorly. There are as many as five of the bifids but only 1–2 of the simple-pointed type. There are thin penis sheaths (newly observed on the type specimen). The species has been reported from several European sites as well as North America.

Tubificoides amplivasatus (Erseus) (Fig. 16*f*,*g*) has a dense, but fine, granulation behind the clitellum, smooth hair setae with bifid setae anteriorly but with simple-pointed setae posteriorly. Thus far it is restricted to Scandinavia, where it was dredged from 70–260 m depth, although specimens, almost certainly of this species, have recently been seen in some Scottish localities.

Tubificoides aculeatus (Cook) (Fig. 16*h*) is an abyssal species from the Northwest Atlantic, and for this and other species in the genus see Brinkhurst and Baker (1979).

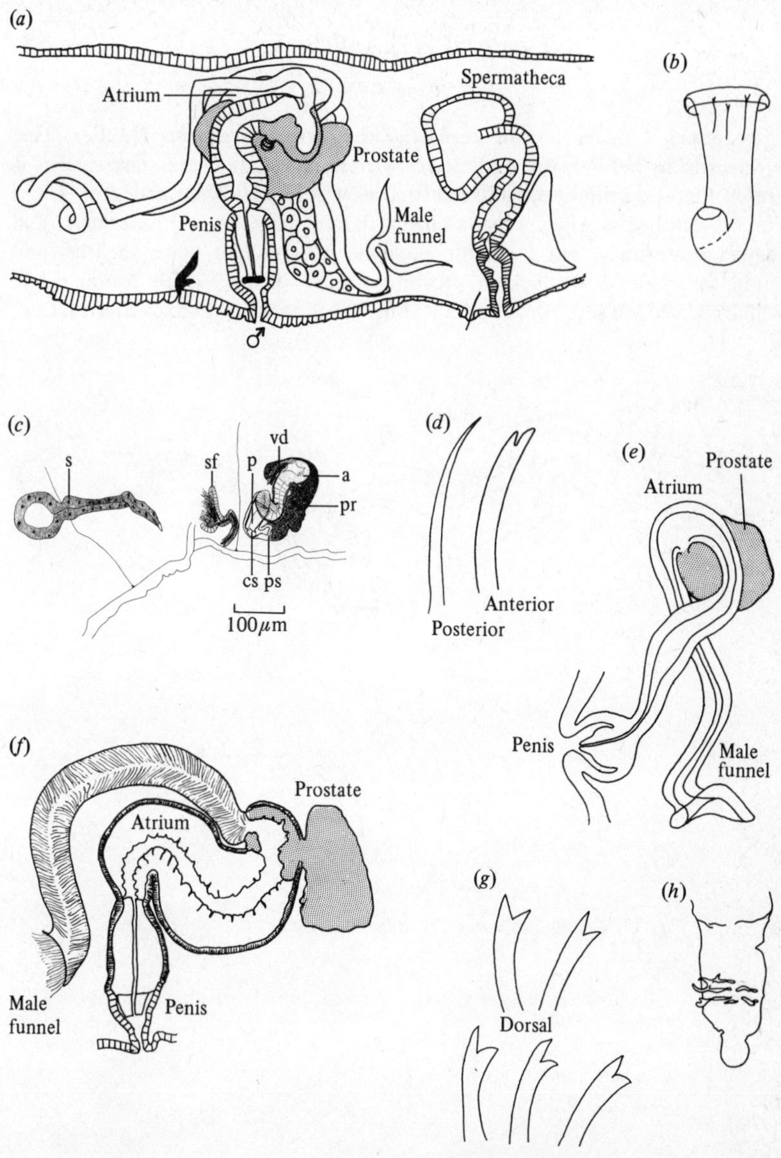

Fig. 16. *Tubificoides* spp.: (*a*) reproductive system and (*b*) penis sheath of *T. benedeni* (after Holmquist, 1978, and Brinkhurst and Jamieson, 1971); (*c*) reproductive system of holotype of *T. pseudogaster* (after Baker, 1980) (key: p = penis; s = spermatheca; sf = sperm funnel; a = atrium; vd = vas deferens; pr = prostate; ps = penial sac; cs = cuticular penis sheath); (*d*) setae and (*e*) reproductive system of *T. heterochaetus* (after Michaelsen, 1926); (*f*) reproductive system and (*g*) setae of *T. amplivasatus* (after Erseus, 1976*b*); and (*h*) penis sheath of *T. aculeatus* (after Cook, 1970*c*).

Genus CLITELLIO Savigny

(Fig. 17)

The genus *Clitellio* is now restricted to *Clitellio arenarius* (Müller) (see *Heterodrilus* below), the earliest known marine oligochaete. The species is often found on the beach and in estuaries with *Tubificoides benedeni.* It has faintly bifid setae (Fig. 17*b*), as in the latter, and also 2–3 anteriorly, and fewer posteriorly, but it has no papillae, no cuticular penis sheaths, and elongate male ducts lacking prostate glands. Timm (1970) found single enlarged ventral setae in X and XII, but I have not been able to confirm this.

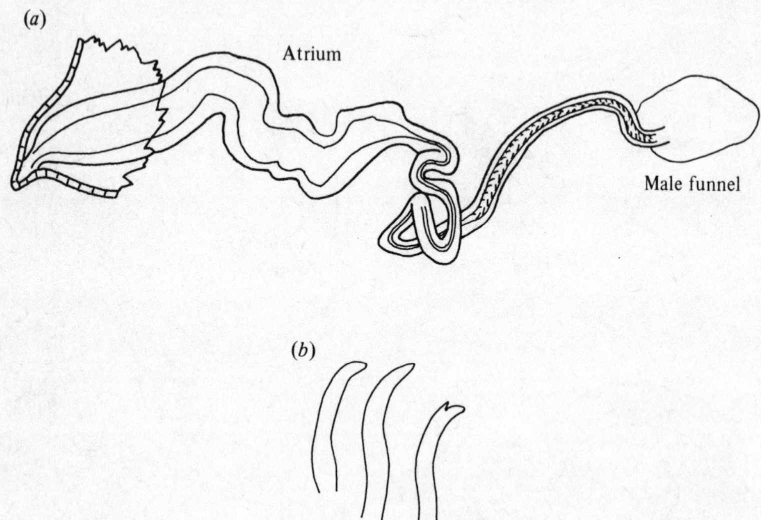

Fig. 17. *Clitellio arenarius*: (*a*) reproductive system and (*b*) setae.

Subfamily AULODRILINAE

Coelomocytes absent, or at least not large and abundant. Prostate glands single on each atrium, broadly stalked or rarely absent. Genital setae as spoon-shaped penials in ventral bundles of XI, or hollow spermathecal setae in X or in X and XI, commonly absent. Pseudopenes present, or small penis-like structures; no thickened cuticular sheaths.

The genus *Thalassodrilides* was erected by Brinkhurst and Baker (1979) and has been shown by C. Erseus (personal communication) to belong to this subfamily along with another marine genus *Limnodriloides* and the freshwater *Aulodrilus* and *Siolidrilus* of the original grouping.

Genus LIMNODRILOIDES Pierantoni

(Fig. 18)

This genus is restricted to species bearing blind-ending diverticula on each side of the intestine in segment VIII or IX, and with male ducts of a characteristic pattern. The large prostate glands on the atria are attached by broadened stalks, typical of the subfamily Aulodrilinae (*Aulodrilus,* possibly *Siolidrilus*, and *Limnodriloides*) and the male ducts usually end in some form of penial structure, though this is never as fully developed as in the Tubificinae.

There are some 10 species, only one of which – *Limnodriloides winckelmanni* Michaelsen, Fig. 18 – is at all widely distributed, being known from Southwest Africa, Mexico (as *L. barnardi* Cook), Scandinavia and Australia. The other species described to date are *Limnodriloides victoriensis* Brinkhurst and Baker (Pacific Canada), *Limnodriloides medioporus* Cook (Massachusetts, USA, and the Northwest Pacific), *Limnodriloides verrucosus* Cook and *Limnodriloides monothecus* Cook (Pacific North America from Mexico to Canada), *Limnodriloides maslinicensis* Hrabe and *Limnodriloides pierantonii* Hrabe (Adriatic), *Limnodriloides fragosus* Finogenova and *Limnodriloides agnes* Hrabe (Black Sea) and *Limnodriloides appendiculatus* Pierantoni (Gulf of Naples). There are other species in the process of being described (C. Erseus, personal communication). Jamieson (1977) placed *L. barnardi* Cook in synonymy with *L. winckelmanni*, both having modified spermathecal setae. The more recently discovered *L. victoriensis* also has genital setae (spermathecal or penial or both) combined with median fused spermathecal and male pores. The male pores open into a median chamber in *L. medioporus. L. verrucosus* is papillate, unlike any other species, and *L. fragosus* has secretory material on the body wall plus foreign matter. *L. appendiculatus* has the gut diverticula in VIII rather than the usual IX of all others bar some specimens of *L. winckelmanni*. Specimens of the former, according to the identification made by W. Boldt in 1926, were collected in Naples, and were deposited in the Zoological Museum, University of Hamburg (No. V 10133). They have the gut diverticula in IX as usual.

The other differences relate to changes in the already small number of setae, and the form of the male ducts, especially their penial structures. Hence *L. agnes* has large penial sacs or eversible pseudopenes, *L. pierantonii* has short conical penes (20 μm long); those of *L. maslinicensis* are 35 μm long. The last two species were described by Hrabe (1971*a*) who placed them in the new genus *Bohadschia* (along with *L. medioporus* Cook), distinguished from *Limnodriloides* chiefly by the presence of 'true' penes.

Other species at one time associated with this genus are *gurwitschi* (see *Thalassodrilides*) and the dubious entities *roseus* and *pectinatus*, both described by Pierantoni from Naples and formerly catalogued with *Spiridion* (q.v.). Hrabe (1967) had briefly allied what is now *Phallodrilus prostatus* with the genus, but later supported the creation of *Thalassodrilus* for the species, which I now regard as a *Phallodrilus* following Erseus's (1975) description of its second pair of prostate glands. *Limnodriloides dniprobugensis* Jarosenko is properly identified as *Potamothrix caspicus* (Lastockin).

None of these species has been recorded from Britain, but there is every reason to believe that some will eventually be discovered.

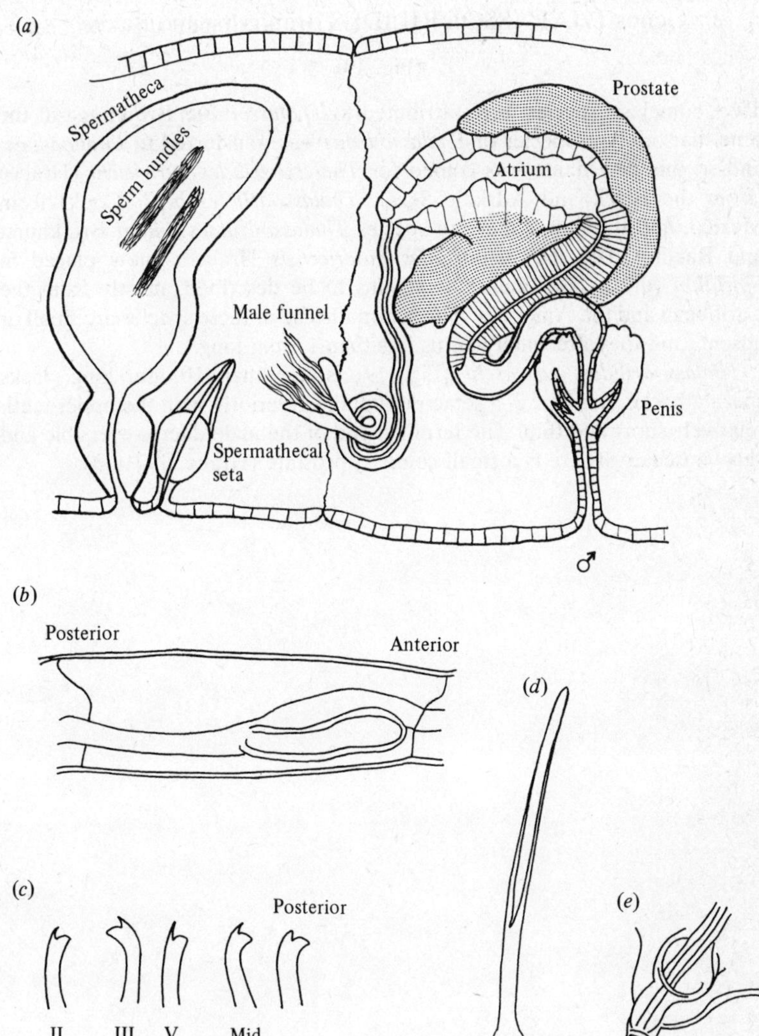

Fig. 18. *Limnodriloides winckelmanni*: (*a*) reproductive organs (after Cook, 1974, as '*barnardi*'); (*b*) gut diverticulum; (*c*) setae; (*d*) tip of spermathecal seta; and (*e*) penes (*b–e* after Jamieson, 1977).

Genus THALASSODRILIDES Brinkhurst and Baker*

(Fig. 19)

Two 'homeless' species were attributed to *Thalassodrilus* Brinkhurst at the time that the type species of *Thalassodrilus* was transferred to *Phallodrilus*, and so this new name was coined for *Thalassodrilides gurwitschi* (Hrabe) (from the Black and Adriatic Seas), *Thalassodrilides belli* (Cook) from Mexico, and the new American species *Thalassodrilides milleri* Brinkhurst and Baker, 1979. *Thalassodrilides modricensis* Hrabe is now placed in *Spiridion* (q.v.). New material remains to be described, mostly from the Caribbean and the Americas. The spermathecae of these species are small or absent, and they are small worms less than 17 mm long.

Thalassodrilides gurwitschi (Fig. 19) is less than 10 mm long, lacks spermathecae, has 2 or 3–5 setae per bundle anteriorly with the upper teeth relatively short and thin. The terminal end of the male duct is eversible and very muscular. There is a small compact prostate (Hrabe, 1971*a*,*b*).

* The name *Curacaodrilus* is a synonym of *Thalassodrilides* having been published (Righi and Kanner, 1979) after the paper by Brinkhurst and Baker (1979).

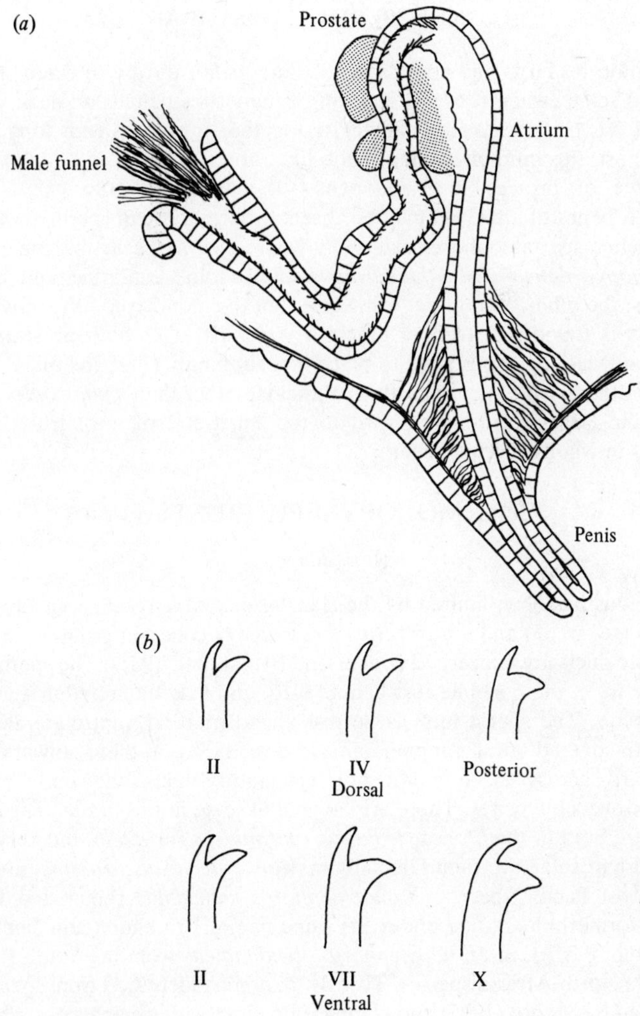

Fig. 19. *Thalassodrilides gurwitschi*: (*a*) reproductive system and (*b*) setae (after Hrabe, 1971*a*).

Subfamily RHYACODRILINAE

Coelomocytes large and abundant. Prostate glands diffuse or rarely absent. Genital setae (when present) in form of numerous penials on each ventral side of XI, clustered with heads close together, proximal ends long, distal ends short, tips bifid or swollen, knob-like, or simple-pointed.* Voluminous eversible or protrusible pseudopenes often present, true penes rarely present, penes of any form may be absent. Sperm loose in spermathecae, or in attached spermatophores externally where spermathecae missing.

Monopylephorus and *Rhizodrilus* contain some estuarine and coastal species, the other genera are freshwater; but the monotypic *Jolydrilus jaulus* Marcus is found in Brazilian brackish water. It lacks both prostates and spermathecae, and so is hard to place in a subfamily (Fig. 20*d*,*e*).

As in the marine species of the Tubificinae other than *Tubificoides*, these are quite large worms in contrast to the small size of most truly marine species in wholly marine genera.

Genus MONOPYLEPHORUS Levinson

(Fig. 20*a–c*,*f*)

This genus has been limited by the transference of several species to other genera (see p. 65) and it now forms a reasonably coherent group in so far as the male ducts are concerned (Baker and Brinkhurst, 1981). The spermathecae are in X, the pores lie just behind 9/10, and may be united in a median depression. The sperm funnels lie just ahead of 10/11 ventrolaterally, the prostate-covered atria running almost directly from them upwards and backwards across XI, to end in short ejaculatory ducts followed by protrusible or eversible penes. There are no genital setae in this group. The dorsal setae are bifid in the *Monopylephorus rubroniveus* subgroup, but very thin, twisted hair setae are found dorsally in *Monopylephorus irroratus*, in a new Northwest Pacific species (*Monopylephorus cuticulatus* Baker and Brinkhurst) formerly identified under the same name (Brinkhurst and Jamieson, 1971: Fig. 8.34*b*), in *Monopylephorus aucklandicus* from the South Pacific, and in a North African species. True *M. irroratus*, identified from specimens collected by Moore (1905) and compared by him with originals described by Verrill from a similar New England locality, have eversible penes with 1–3 cuticular processes at one spot on each inner wall. These were mistakenly labelled as penes of *M. rubroniveus* (Brinkhurst and Jamieson, 1971: Fig. 8.35*e*), and it should be noted that the hair setae can very easily be broken off, which creates some of this confusion. This species has been recorded from Wimereux, France, and is the same as a specimen identified as *Monopylephorus trichochaetus* Ditlevsen by W. Boldt (specimen in the Hamburg University collection) which may be a type of the synonymous

* Plus pre-spermathecal setae in the American freshwater species *Rhizodrilus lacteus* and in the marine *R. pacificus*.

Postiodrilus sonderi Boldt. Another species with similar hair setae but with thin, coiled, tubular, cuticular penial linings in XI is known from the River Weaver in Cheshire. *Monopylephorus parvus* Ditlevsen has simple-pointed setae posteriorly, bifids anteriorly, male ducts like those of the rest of this group (see below) but has a single spermatheca with a mid-ventral pore.

The original brief description of *M. rubroniveus* Levinsen was augmented by Ditlevsen, but there has always been the possibility that he was not looking at the same species and so some authors have turned to *Rhizodrilus lacteus* as the generic type, preferring this generic name (Hrabe, 1967, for example). Recent work (Baker and Brinkhurst, 1981) shows that this will not suffice as *Rhizodrilus* (see p. 68) is distinctly different to *Monopylephorus:* it is not recorded from Europe. However, the very large, median, male copulatory bursa, simple protrusible penes, lack of hair and simple-pointed setae, and closely paired or united spermathecal pores in a median depression close behind 9/10 are unique to *M. rubroniveus*. A number of similar forms has been described from different localities, but these are all regarded as synonymous. None of them is well enough described or from a locality that would substantiate the possibility that Ditlevsen failed to collect Levinsen's original species at the type locality.

Specimens of these species can be found inland, as evidenced by the discovery of *M. rubroniveus* in the River Tame by R. W. Martin (confirmed by me) and *M. irroratus* in the Irwell (Eyres, Williams and Pugh-Thomas, 1978) but the identity of the latter may have been based solely on the presence of twisted hair setae, which will no longer suffice. My own '*M. irroratus*' from the River Weaver is a distinct species which cannot be described until new material is obtained.

Several former species of *Monopylephorus* are now included in *Rhizodrilus* (see p. 68), whilst *M. montanus* is currently placed in a new genus *Peristodrilus*, and a second new genus is being created for *M. frigidus*, *M. longisetosus* and some as yet undescribed species. None of these is likely to occur in Britain.

The four potentially British species can be identified as follows although British records will require confirmation apart from *M. rubroniveus* from the River Tame.

Monopylephorus parvus Ditlevsen

(Fig. 20*f*)

Small worms (8–15 mm) with bifid setae anteriorly, simple-pointed posteriorly. Single spermatheca with median pore. Male pore median, with protrusible penes.

Monopylephorus rubroniveus Levinsen

(Fig. 20*a*)

Setae bifid. Spermathecal pores closely paired or opening to median depression. Male pore median, with protrusible penes.

Monopylephorus irroratus (Verrill)

(Fig. 20*b*,*c*)

Dorsal setae (?) bifid to pectinate accompanied by thin, twisted hair setae dorsally (watch for broken stumps). Male and spermathecal pores paired, ventrolateral, penes eversible with cuticular processes on linings.

(Unnamed species from River Weaver: dorsal setae pectinate accompanied by thin, twisted hair setae dorsally. Penes presumably eversible, coiled, with thin cuticular linings.)

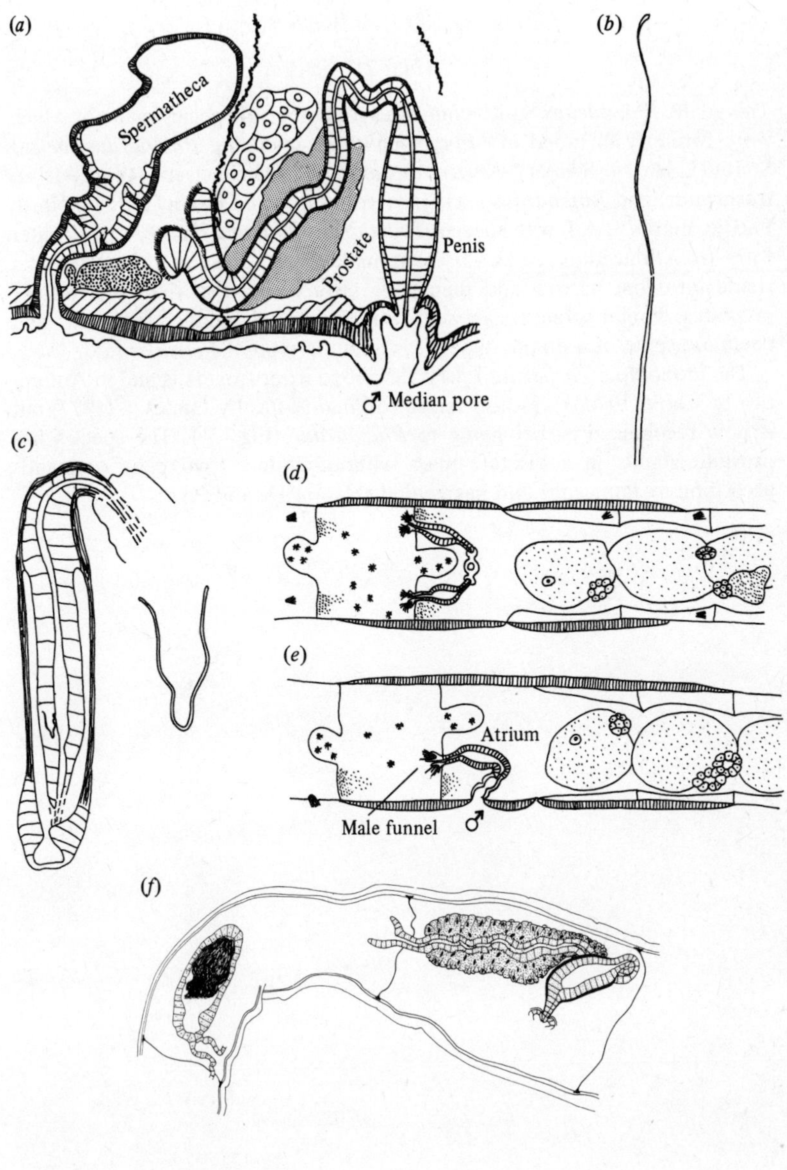

Fig. 20. *Monopylephorus* and *Jolydrilus*: (*a*) reproductive system of *M. rubroniveus*; (*b*) hair seta of *M. irroratus*; (*c*) penis of *M. irroratus* with an enlargement of the cuticular process from the lining of the penis (after Brinkhurst and Jamieson, 1971); (*d*, *e*) reproductive system of *Jolydrilus* (after Marcus, 1965); (*f*) reproductive system of *M. parvus* (after Baker and Brinkhurst, 1981).

Genus RHIZODRILUS Smith

(Fig. 21)

The genus *Rhizodrilus* Smith comprises a number of species many of which were formerly included in *Monopylephorus*, including *Rhizodrilus lacteus* Smith (USA, freshwater), *Rhizodrilus africanus* (Michaelsen) (West Africa, freshwater) and *Rhizodrilus pacificus* (Brinkhurst and Baker) (circum North Pacific, marine). A fourth species, *Rhizodrilus montana* Hrabe, a freshwater form from Macedonia, has very long male ducts, and the vasa deferentia shade into first narrow and then wide atria covered in the usual diffuse prostates. Penial setae are present, but the dorsal bundles have hair and pectinate setae of a unique type. It is now the type of *Peristodrilus*.

The monotypic *Torodrilus lowryi* described from Anvers Island in Antarctica by Cook (1970*b*) was transferred to *Phallodrilus* by Jamieson (1977) but is now recognized as belonging to *Rhizodrilus* (Fig. 21). The species has prostate glands in a discrete mass without a duct (two pairs of glands according to Jamieson) and has genital setae in IX and XI.

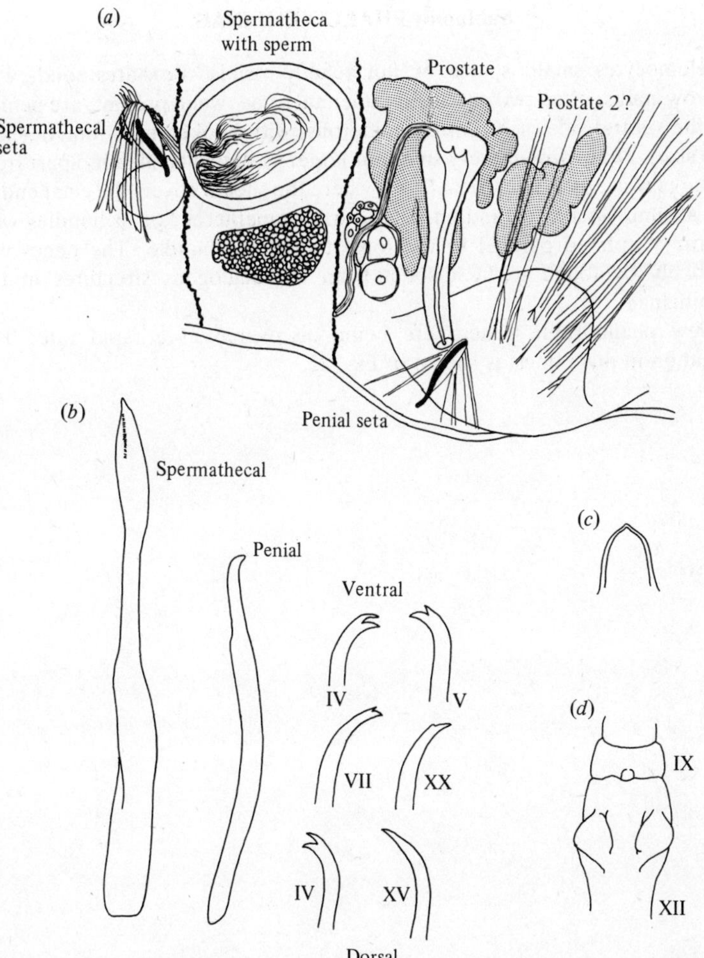

Fig. 21. *Rhizodrilus lowryi*: (*a*) reproductive system; (*b*) setae; (*c*) shape of prostomium; and (*d*) ventral view of segments IX–XII (after Cook, 1970*b*).

Subfamily PHALLODRILINAE

Coelomocytes small, sparse or not readily visible. Prostates solid, with narrow stalks, often two per atrium. Genital setae, when present, are penials on the ventral side of XI, often numerous, with hooked tips, sometimes in two sizes. Penes absent, very small, or large; no cuticular sheaths apart from in the new genus *Bermudrilus* (q.v.) where the sheath covers the ectal end of the atrium. Sperm in random masses in spermathecae, or in bundles with sperm orientated parallel to the long axis of the bundle. The penes will probably be found to be distinct from the analogous structures in the Tubificinae.

New phallodriline genera are being discovered at a rapid rate. The variation in male ducts is shown in Fig. 22.

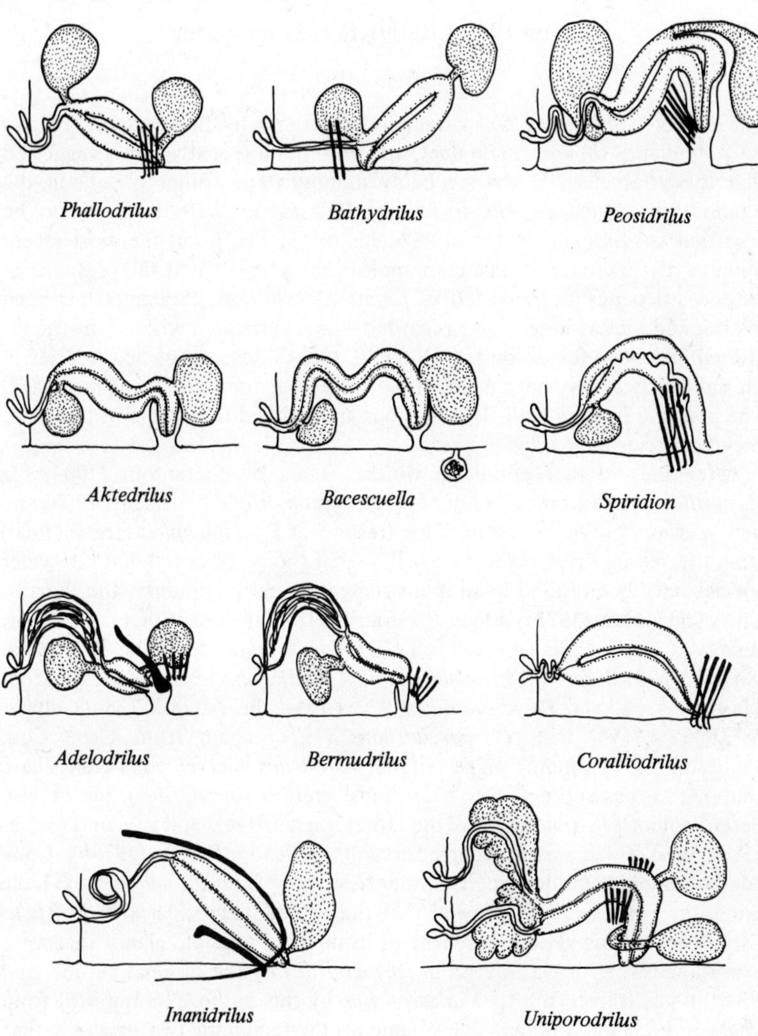

Fig. 22. Reproductive systems and male ducts of phallodriline genera (after Erseus, 1980*a*).

Genus PHALLODRILUS Pierantoni

(Fig. 23)

The genus *Phallodrilus* was originally diagnosed by the presence of two prostate glands on each male duct, and until quite recently it was suggested that this characteristic was probably monophyletic, other genera in the Phallodrilinae such as *Aktedrilus* and *Bacescuella* therefore came to be regarded as synonyms (Cook, 1969*b*; Jamieson, 1977), but the most recent opinion (C. Erseus, personal communication) suggests that the presence or absence of penes be regarded as a generic criterion, though it has been overlooked elsewhere or regarded as variable within a genus (*Limnodriloides*, for example). With the rapid discovery of new species in the subfamily it may be sensible to be somewhat conservative at this stage. The penes appear to be little more than unmodified terminalia of the atria. Eleven species lack a gut.

After the original definition of the genus by Pierantoni (1902) for *P. parthenopaeus* from the Gulf of Naples (redescribed by Erseus, 1980*b*) no new species were described until the freshwater *P. aquaedulcis* Hrabe (1960) from the Weser River in Germany (though it should be noted that salt-water species have been found in an industrially influenced tributary, the Werra). The same author (1971*a*) added *P. adriaticus* from the sea of that name – now seen as a *Bathydrilus* (q.v.) – and (1973) *P. minutus* from the Black Sea. Cook (1969*b*) described *P. coeloprostatus* and *P. obscurus* from Cape Cod, Massachusetts, and *P. profundus* (1970*a*) from the Atlantic (38°08′–46′N, 68°32′–71°47′W), with *P. parviatriatus* (1971) again from Cape Cod. Jamieson (1977) not only suggested that *Torodrilus lowryi* Cook belongs here (although it has subsequently been transferred to *Rhizodrilus*), but he also described both *P. albidus* (from the Great Barrier Reef) and a second species (*P. rohdei*) which was transferred to *Bathydrilus* by Erseus (1979*b*). Cook and Hiltunen (1975) described another freshwater form, *P. hallae*, from Lake Superior. Erseus (1975) showed that the problematic species *P. prostatus* (Knöllner) belonged here by virtue of its doubled prostate glands that were overlooked when it was placed in *Rhyacodrilus* by the original author, and when it was transferred to *Thalassodrilus* by this author. Its removal from *Thalassodrilus* necessitated a new name for the remaining two species in that genus once the type species was removed (see *Thalassodrilides*). Other new species have now been found in North America, Australia, and the Comoro Islands (Erseus, 1979*f*, 1981).

Only the type species is definitely known from Britain, *P. parthenopaeus* being found in the Outer Hebrides (Erseus, 1980*b*). Few of the existing species are obvious candidates for the British list as they are each known from a single location, with the exception of *P. prostatus,* reported by

Brinkhurst (1963*b*) from the Menai Straits, but which requires confirmation in view of the diversity now recognized.

Phallodrilus prostatus (Knöllner)

(Fig. 23*a*)

6–10 mm long, with 3–7 (or more or less) dorsal and 5–9 ventral bifid setae anteriorly, posteriorly 4–6 or 7 simple-pointed setae, with as many as 22 single-pointed penial setae on each side of XI. The atria are short, broad and muscular; there are no penes (Fig. 23*a*). The most recent account is that of Erseus (1975).

Phallodrilus parthenopaeus Pierantoni

(Fig. 23*b,c*)

(From Scotland) 3.1 mm long; setae 2–3 per bundle anteriorly, only one posteriorly. The ventral setae of X are paired, slightly elongate and broad, with long curved upper teeth; the penials of XI are twice as long as the ventrals, very wide basally, narrow apically (Fig. 23*b,c*).

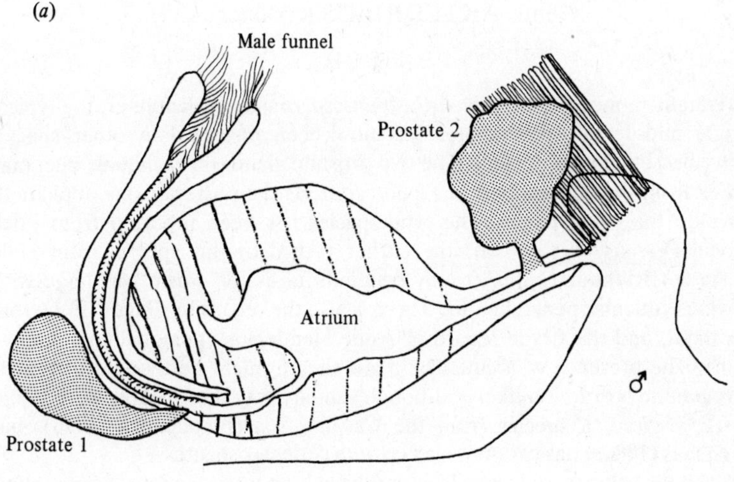

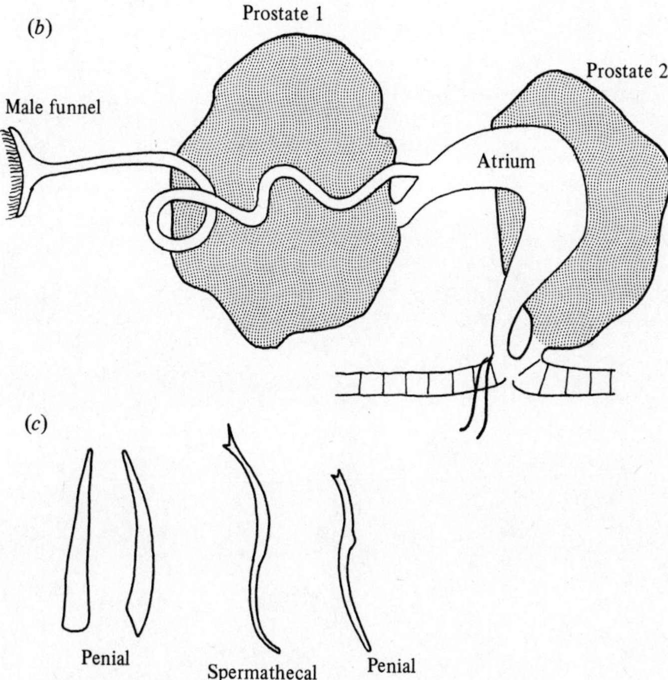

Fig. 23. *Phallodrilus* spp.: (*a*) reproductive system of *P. prostatus* (after Erseus, 1979*f*); (*b*) reproductive system and (*c*) setae of *P. parthenopaeus* (after Pierantoni, 1902).

Genus AKTEDRILUS Knöllner 1935

(Fig. 24)

Originally monotypic for *Aktedrilus monospermathecus* Knöllner, the typical single mid-dorsal spermatheca has now been observed in other species described by Erseus (1980*c*). The two prostate glands on each male duct may differ from those described in *Phallodrilus*, as the posterior one appears to envelop the small penes. The type species has been reported from Loch Leven (Erseus, 1980*c*), but the earlier record of this species from Hale (Lancs) (Brinkhurst, 1963*a*) may be doubtful as the worm was large with obvious cuticular penis sheaths. There is also the record by Gage (1974) from Scotland, and the Clyde record of Anne Henderson (personal communication). The presence of a cuticular penis sheath might be thought to render this generic position unlikely, although similar, very thin sheaths were found in *A. svetlovi*, a species from the Caspian Sea (Finogenova, 1976), but C. Erseus (1980*c*) has two new species with cuticular sheaths.

Even the solitary mid-dorsal spermatheca is no longer unique, being found in a species of *Bacescuella* (p. 84) and in *Inanidrilus bulbosus* Erseus (Erseus, 1979*g*).

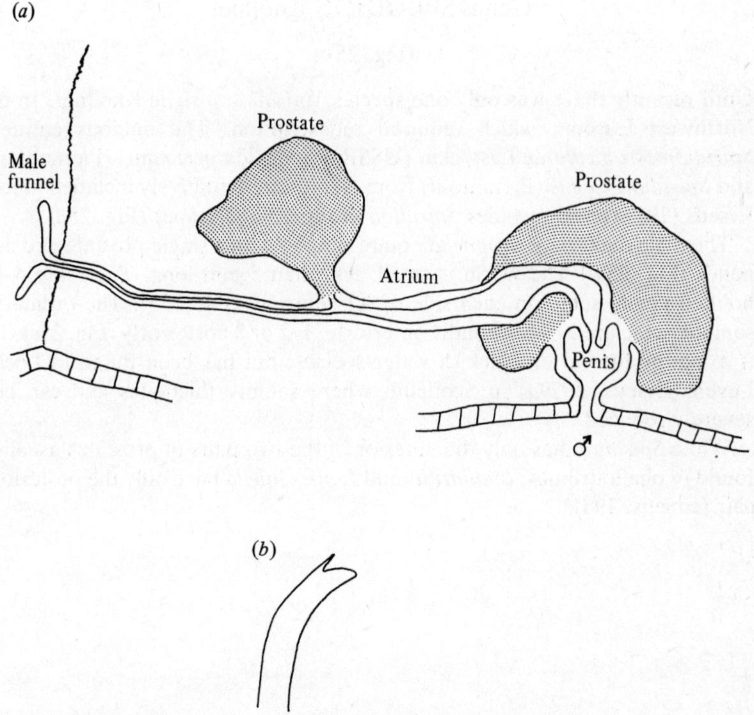

Fig. 24. *Aktedrilus monospermathecus*: (*a*) reproductive system and (*b*) seta (after Knöllner, 1935).

Genus SPIRIDION Knöllner

(Fig. 25)

Until recently there was only one species, *Spiridion insigne* Knöllner, from Northwest Europe, which required redescription. The dubious entities *Spiridion scrobicularae* Lastockin (USSR), *Spiridion pectinatus* (Pierantoni) and *Spiridion roseus* (Pierantoni) from Naples are tentatively included here. Erseus (1979*a*) also includes *Spiridion modricensis* (Hrabe) (Fig. 25*d*).

The male ducts of *S. insigne* are quite simple, with a single prostate and no penes (Fig. 25*a*). The worm is small, less than 5 mm long. There are 4–6 hooked penial setae on each side of XI ventrally (Fig. 25*b*). The ordinary somatic setae are 3–5 per bundle anteriorly, 1–2 or 3 posteriorly (Fig. 25*c*). It is a marine littoral or brackish water species, but has been found in Loch Leven (Erseus, 1979*a*) in Scotland, where salinity fluctuates and can be severely reduced.

While *Spiridion* has only the anterior of the two pairs of prostates usually found in phallodrilines, *Inanidrilus* and *Jamiesoniella* have only the posterior pair (Erseus, 1981).

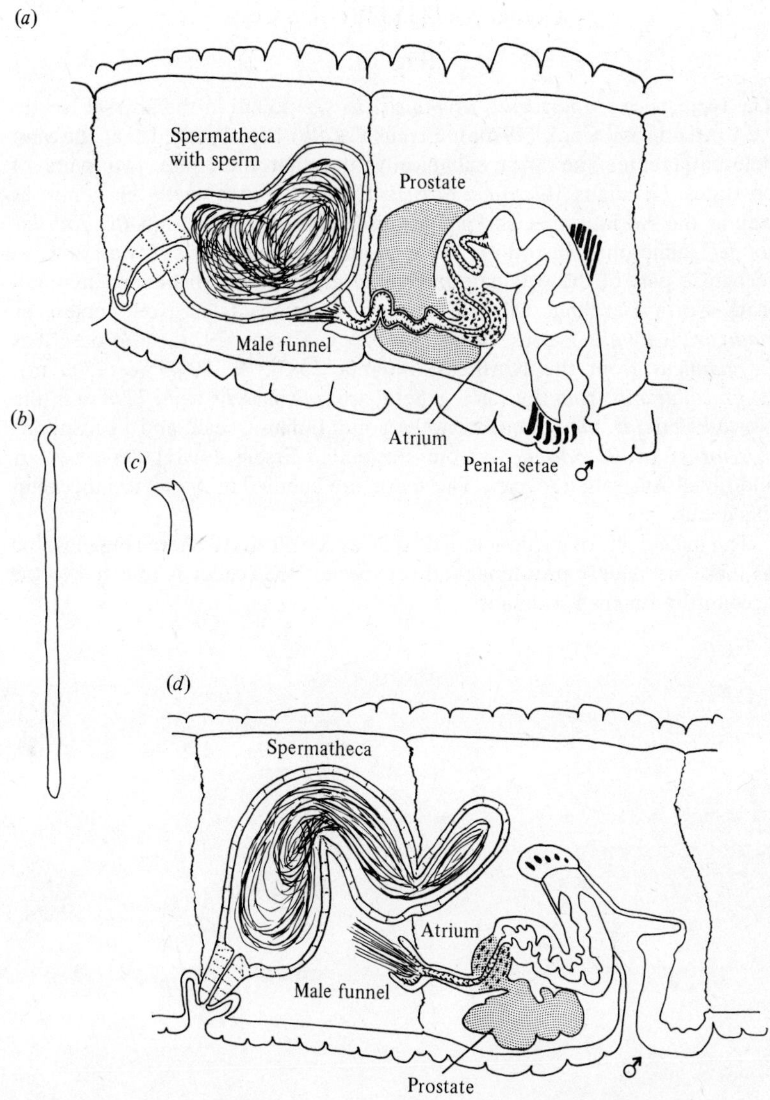

Fig. 25. *Spiridion* spp.: (*a*) reproductive system, (*b*) penial seta and (*c*) seta of *S. insigne*; (*d*) reproductive system of *S. modricensis* (after Erseus, 1979*a*).

Genus BATHYDRILUS Cook

(Fig. 26)

The type species *Bathydrilus asymmetricus* was found in the abyssal Northwest Atlantic by Cook (1970*a*). Erseus (1979*b*) has observed that the vasa deferentia enter the atria subapically and that there are two pairs of prostates. He aligns *Macroseta rarisetis* Erseus (Norway) with this genus, as well as the Adriatic species *Phallodrilus adriaticus* Hrabe, and *Phallodrilus rohdei* Jamieson (Australia). As a result of seeing this, I was able to recognize part of the original *Smithsonidrilus marinus* Brinkhurst material from North Carolina, as well as new specimens from New Jersey, as *Bathydrilus longus* Erseus. The same paper (Erseus, 1979*b*) also describes *B. atlanticus* from the Northeast Atlantic (58°42′N, 09°43′W, 1800 m), *B. graciliatriatus* from the same general area, *B. hadalis* from 7298 m in the North Pacific, *B. meridianus* from the South Indian Ocean, and a subspecies (*trisetosus*) of *B. adriaticus* from Bermuda. Erseus (1981) describes an additional Australian species. The sperm are bundled in the spermáthecae in this genus.

The male ducts of a characteristic species are illustrated here (Fig. 26), but as these are mainly profundal/bathyal species, the reader is referred to the account by Erseus for details.

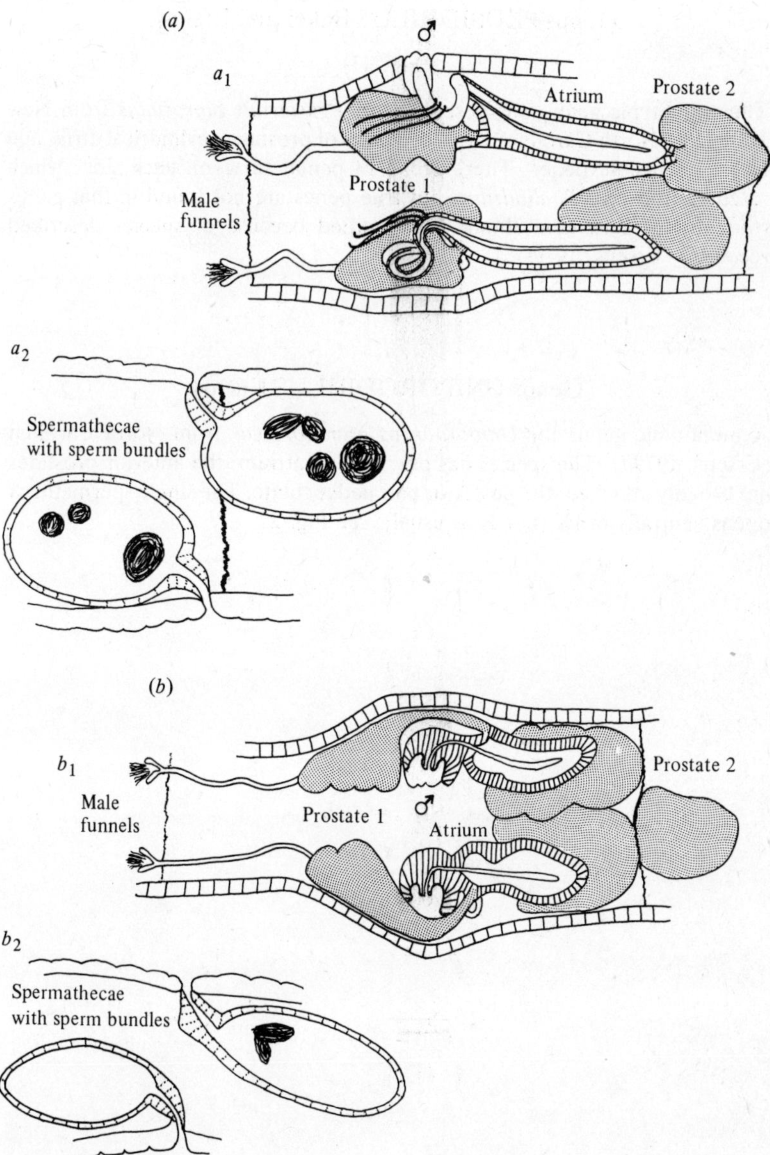

Fig. 26. *Bathydrilus* spp.: reproductive system and spermathecae with sperm bundles of *B. graciliatriatus* (a_1 and a_2) and of *B. asymmetricus* (b_1 and b_2) (after Erseus, 1979*b*).

Genus PEOSIDRILUS Baker and Erseus

(Fig. 27)

This monotypic genus was described for *Peosidrilus biprostatus* from New Jersey and North Carolina. It has two pairs of prostates, cylindrical atria, and large protrusible penes. There are 7–13 penial setae on each side, which resemble those of *Phallodrilus*, but true penes are not found in that genus (though this statement should be modified because of species described recently – Erseus, 1979*f*).

Genus UNIPORODRILUS Erseus

A monotypic genus for *Uniporodrilus granulothecus* from North Carolina (Erseus, 1979*d*). The species has one median atrium, the anterior prostates are broadly attached, the posterior pair pedunculate. The single spermatheca opens ventrally in IX (not X as usual); see Fig. 22.

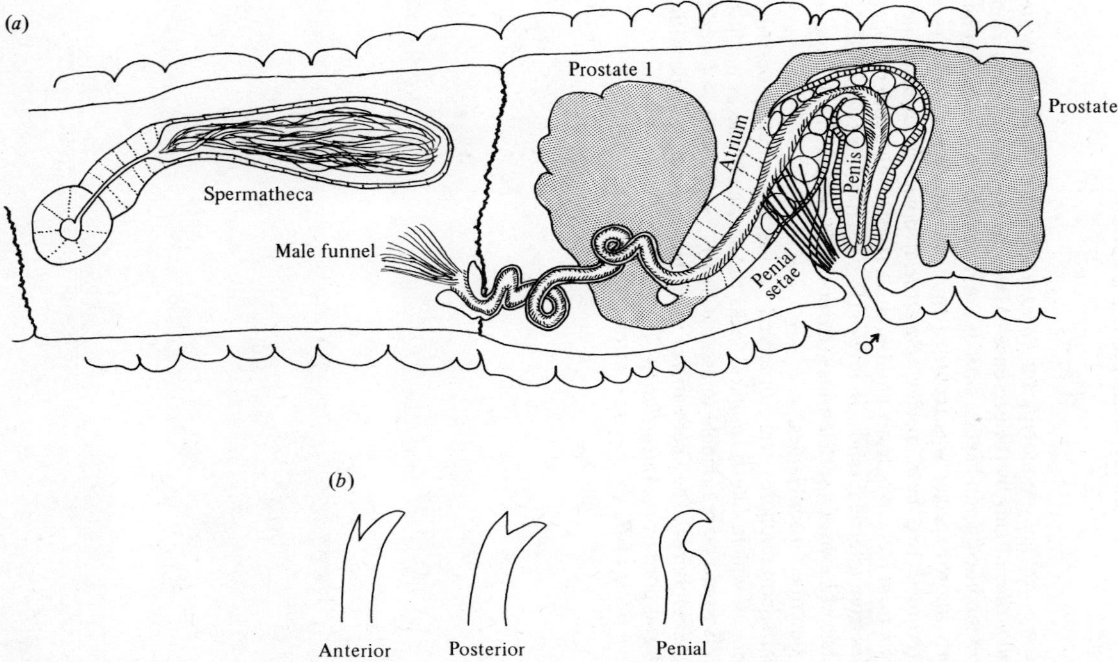

Fig. 27. *Peosidrilus biprostatus*: (*a*) reproductive system and (*b*) setae (after Baker and Erseus, 1980).

Genus BACESCUELLA Hrabe

(Fig. 28)

The genus was originally monotypic for *Bacescuella pontica* Hrabe, separated from *Phallodrilus* largely by the presence of large penes in penis sacs. There are also no modified genital setae, such as are usually present in that genus. This poorly known species was recorded from the Black Sea, but Erseus (1978*a*) recorded two new species, *Bacescuella arctica* and *Bacescuella parvithecata* from Iceland/Norway and Bermuda respectively; both meiobenthic species from intertidal sands. *B. arctica* (Fig. 28) has now been recorded from the Isle of Lewis, Outer Hebrides (Erseus, 1980*c*), and a fourth species has been described from Naples.

External spermatophores were observed on the two recently described species, one of which, like *B. pontica*, lacks the spermathecae as might be expected. It would be reasonable to assume that *B. pontica* will be found to carry spermatophores. The spermatheca in *B. parvithecata* is single with a median dorsal pore as in *Aktedrilus* (p. 76).

The male ducts of *B. arctica* are illustrated here.

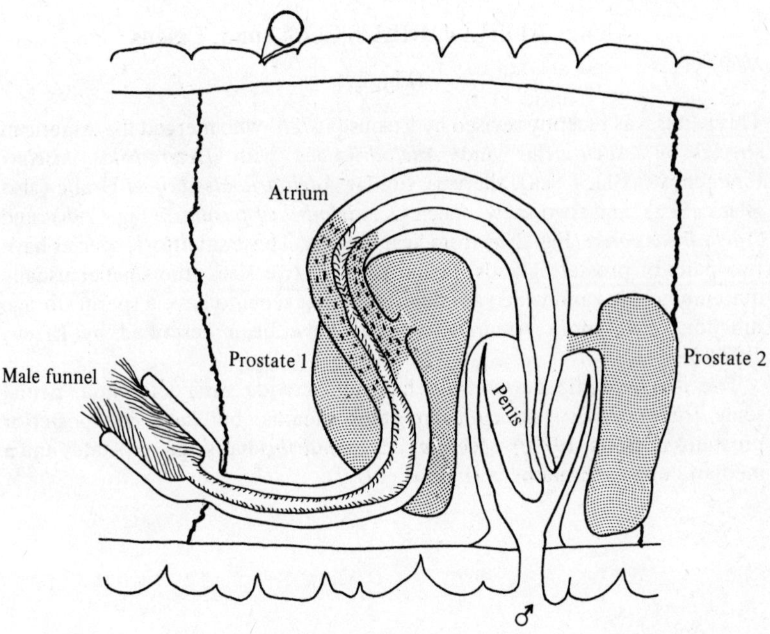

Fig. 28. Reproductive system of *Bacescuella arctica* (after Erseus, 1980*d*).

Genus ADELODRILUS Cook emm. Erseus

(Fig. 29)

This genus was recently revised by Erseus (1978*b*) who merged the American species of *Adelodrilus* and *Adelodriloides* with *Heterodrilus kiselevi* Finogenova (Black Sea), the very similar *Adelodriloides borcei* Hrabe (also Black Sea) and two new species, *Adelodrilus pusillus* (Fig. 29*b*) and *Adelodrilus cooki* (Fig. 29*a*) from Scandinavia. These sublittoral species have two pairs of prostate glands, penial setae of two sizes (the smaller usually quite numerous) and wide vasa deferentia that seem to have a sperm storage function. Two more American species have been described by Erseus (1979*e*).

The monospecific *Bermudrilus* has similar wide vasa deferentia, penial setae (of one size) and cuticular penis sheaths, but lacks the posterior prostates (Erseus, 1979*e*) – see Fig. 22. *Inanidrilus* has single prostates and a median dorsal spermatheca (Erseus, 1979*f*).

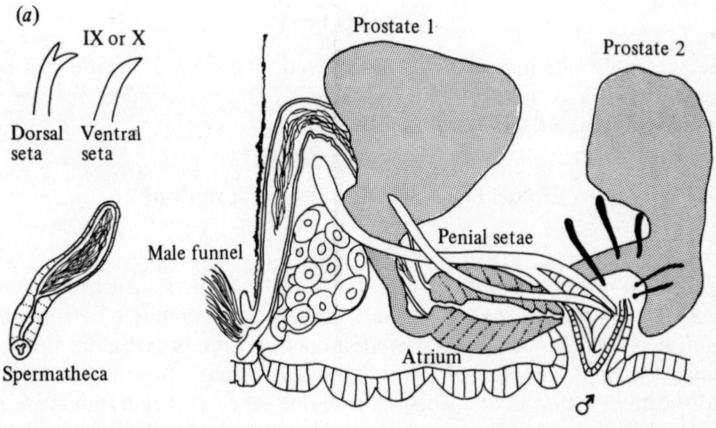

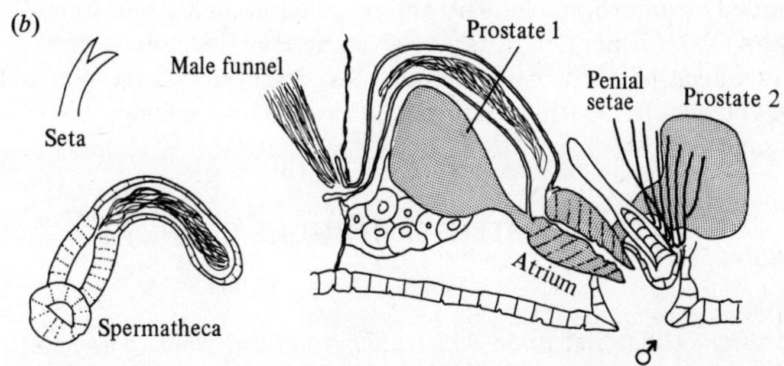

Fig. 29. *Adelodrilus* spp.: (*a*) *A. cooki*; (*b*) *A. pusillus* (after Erseus, 1978*b*).

Other genera

The subfamily Clitellinae has been dissolved, and *Clitellio arenarius* is now recognized as a member of the Tubificinae. This leaves some species not currently assignable to any subfamily.

Genus HETERODRILUS Pierantoni

(Fig. 30*a*)

This is now reinstated to its generic status, and is under review by C. Erseus (personal communication). Originally thought to be unique by virtue of its trifid setae, it now seems that more than one species, in fact more than one genus, has this characteristic. The Northwest Atlantic forms seem to have two prostates rather than a single diffuse one on each atrium, but there is a single prostatic mass in the subspecies *Heterodrilus arenicolus queenslandicus* Jamieson, which resembles the original description of American material by the author (Brinkhurst, 1966*b*; Jamieson, 1977). The original locality is the Bay of Naples, from where Pierantoni described several species now regarded as unidentifiable. The author failed in an attempt to collect fresh material. No further details can be given here as Erseus has yet to complete his investigation of the genus, though his most recent suggestion is that the genus belongs in the Rhyacodrilinae by virtue of its coelomocytes and diffuse prostate.

Genus SMITHSONIDRILUS Brinkhurst

(Fig. 30*b*)

The original material upon which this American genus was based is now regarded as comprising two species. The Florida types can be retained in the genus, but the North Carolina specimens are a species of *Bathydrilus* (p. 80). The sperm are in bundles, not in spermatozeugmata. There are paired oesophageal diverticula as in *Limnodriloides*, broadly attached prostates, and atrial diverticula. It has recently been placed in the Aulodrilinae.

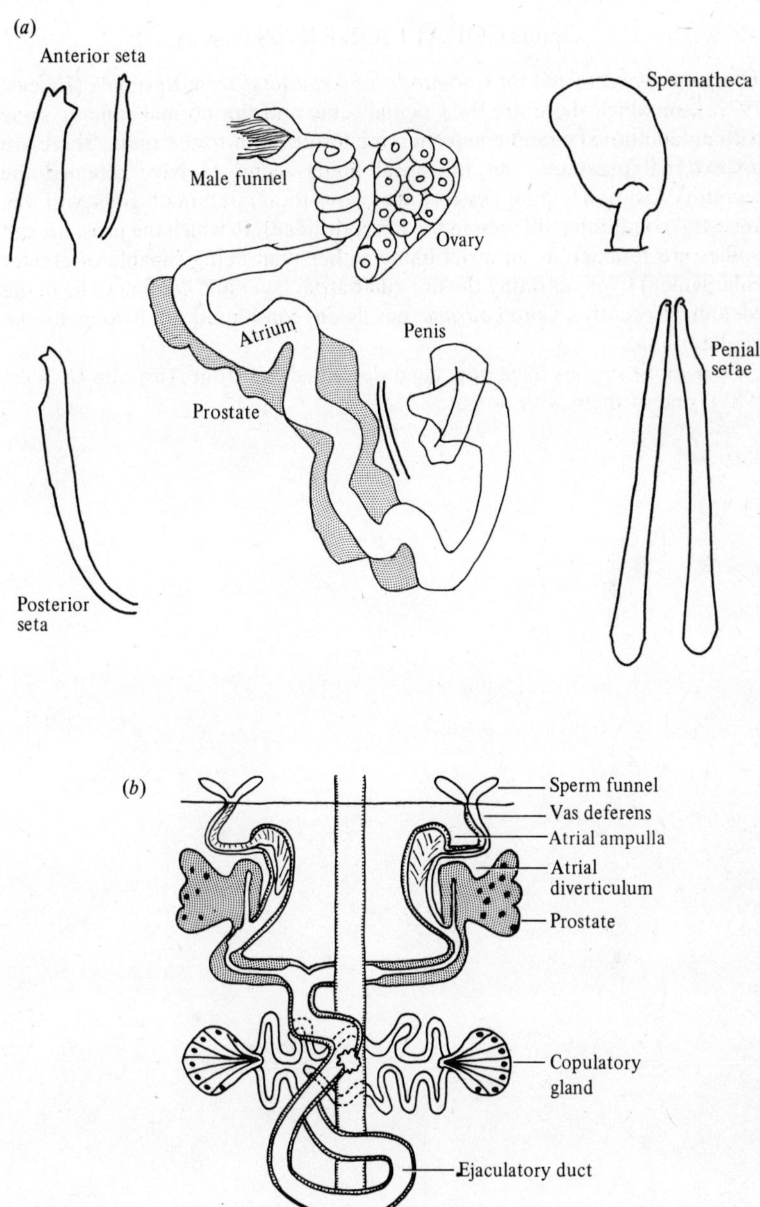

Fig. 30. (*a*) *Heterodrilus ?arenicolus*; (*b*) *Smithsonidrilus marinus*, male duct (after Erseus, in preparation).

Genus CORALLIODRILUS Erseus

This genus was erected for *Coralliodrilus leviatriatus* from Bermuda (Erseus, 1979*c*), in which there are 9–11 penial setae and the normal somatic setae with an additional strand connecting the lower tooth to the shaft. There are no external prostates, but the atrial lining seems to have retained the secretory function – quite possibly analogous to, if not homologous with, the ancestral condition (still seen in the Phreodrilidae), in which the prostatic cell bodies are retained as an atrial lining rather than being capable of greater enlargement by penetrating the muscular atrial layer and coming to lie in the coelom. Recently, *Coralliodrilus* has been considered to belong to the Phallodrilinae.

Four more species have now been described, all from Australia (Erseus, 1981), one of them with no gut.

Family ENCHYTRAEIDAE

There is no modern, widely accepted, subfamily classification of the Enchytraeidae. The species list used here (compiled by Kathryn Coates) includes brackish-water and marine forms found in littoral habitats extending from the emergent vegetation to the lowest intertidal, as well as some subtidal species of *Grania, Lumbricillus* and *Marionina.* Species of these genera (and some *Enchytraeus* species) frequently dominate the enchytraeid fauna of marine habitats, whereas species of *Achaeta, Cernosvitoviella, Cognettia, Fridericia, Hemifridericia, Henlea* and *Mesenchytraeus* are found in small numbers and may be restricted to the highest intertidal, terrestial or freshwater situations.

Guide to the identification of salt-water Enchytraeidae

Decision level 1

A. Setae absent. Two dorsal oesophageal appendages in V. Spermathecal ampullae extending into VIII .... *Achaeta littoralis* Lasserre (p. 94)

B. Setae absent. Transition of oesophagus to intestine gradual. Without oesophageal appendages. Spermathecae contained in V

Marionina arenaria Healy (p. 110)
Marionina achaeta Hagen (p. 110)

C. Setae sigmoid, slender, 2–7 per bundle, with nodulus. Head pore at 0/1. Living worms white, small, 26–28 segments. Ectal duct of spermathecae (from ampulla to body wall) long, narrow and non-glandular; ampullae about 3 times as wide and 1.5 times as long as duct, sac-like, not attached to oesophagus .... *Cernosvitoviella immota* (Knöllner) (p. 94)

D. Setae sigmoid, with nodulus, stout, 1–12 per bundle. Lateral bundles of V–VII with 1–2 setae about 1.5 times longer than others. Head pore at apex of prostomium. Medium-sized, greyish-brownish, 42–55 segments. Ectal duct of spermathecae long and narrow, about twice as long as ampulla, latter attached to oesophagus, single oval to spherical diverticulum arising from ectal end . *Mesenchytraeus armatus* (Levinsen) (p. 94)

E. Setae straight, without nodulus, 2–3 per bundle. Very small worms; 19–21 segments. Two types of coelomocytes present (large oval, granulate and very small, elongate, hyaline). Ental ducts of spermatheca (from ampulla to oesophagus) unite before mid-dorsal connection with oesophagus

Hemifridericia parva Nielsen and Christensen (p. 94)

F. Setae straight or slightly bent, unequal or equal in size in bundle, no nodulus. Oesophageal appendages present in IV, or IV and VI. Postseptal parts of nephridia large and pear-shaped; efferent duct long and narrow, arising posteroventrally near septum. Abrupt transition between oesophagus and intestine. (Typically terrestrial, but may occur in soil at top of shore) .... *Henlea* Michaelsen (p. 98)

G. Setae one per bundle where present, missing from more anterior dorsal than anterior ventral segments .... *Grania* Southern (p. 100)

H. Setae straight, no nodulus, of one size or different sizes in same bundle. A pair of peptonephridia present .... **Level 2(i)** (p. 93)

I. Setae sigmoid or straight, no nodulus, several per bundle. Coelomocytes of one kind. No peptonephridia .... **Level 2(ii)** (p. 93)

Level 2(i)

Species with paired peptonephridia

A. Setae of a bundle arranged in pairs, innermost pairs progressively shorter than outer ones. Dorsal coelomic pores present from about VII. Two kinds of coelomocytes present, large nucleate and small hyaline non-nucleate. Seminal vesicle may be absent . *Fridericia* Michaelsen (p. 102)

B. Setae of a bundle more or less equal – not arranged in pairs. Dorsal coelomic pores absent. Only nucleate coelomocytes present. Seminal vesicles present, usually well developed . . . . *Enchytraeus* Henle (p. 104)

Level 2(ii)

Species with sigmoid or straight setae, no nodulus. No peptonephridia. One kind of coelomocyte.

A. With well-developed, lobed, seminal vesicles
Lumbricillus Oersted (p. 106)

B. Often with more than three pairs of primary pharyngeal glands. Testes and ovaries may be displaced forward by up to four segments. Spermathecae simple, not attached to oesophagus, ectal ducts usually with a single compact gland at ectal orifice. Anteseptale of nephridia comprises funnel only, efferent duct usually arising anteroventrally
Cognettia Nielsen and Christensen (p. 96)

C. Only two or three pairs of primary pharyngeal glands. Spermathecae may have diverticula, usually attached to oesophagus. Glands usually along the ectal duct and/or at ectal orifice. Anteseptal parts of nephridia usually consisting of funnel plus a few coils of nephridial canal, efferent duct normally arising in posterior region of postseptale
Marionina Michaelsen (p. 110)

A. Genera with single marine species

Achaeta littoralis Lasserre

(Fig. 31*a*)

This species, described from Bassin d'Arcachon, France, is found in the upper intertidal, along with *Marionina spicula* (among others) which has been recorded from Britain. The species was described by Lasserre (1968), the key characters being the lack of setae, the abrupt transition of oesophagus into the intestine, many glandular bodies on the penial bulb, and spermathecae not attached to the oesophagus – these being generic characteristics found in this, the only known littoral species.

Cernosvitoviella immota Knöllner

(Fig. 31*b*)

This species has been recorded from Ireland (Healy, 1979*b, c*) and from Anglesey (Tynen, 1966, 1972), from freshwater outlets on the beach and grass above high-water mark. The worm is small (5–7 mm) with pointed, sigmoid setae with a nodulus, 5 or 6 per bundle anteriorly, 2–3 posteriorly.

Mesenchytraeus armatus (Levinsen) (*? Analycus* Levinsen) (=*M. setosus* Michaelsen)

(Fig. 31*c*)

It is widely distributed in Europe, and is essentially a terrestrial species. The setae of the lateral bundles of V–VII with only 1–2 setae, about 1.5 times longer than those of preceding or following segments. The worm is grey-brown, rather flaccid. Brinkhurst and Jamieson (1971) have suggested that nomenclatural rules require this genus to be named *Analycus* Levinsen.

Hemifridericia parva Nielsen and Christensen

(Fig. 31*d*)

This 2–3 mm long species is recorded from Denmark and the Netherlands, but only once from the shore, and it is basically terrestrial (Tynen and Nurminen, 1969). The genus is monotypic. The species has straight, non-nodulate setae, 2–3 per bundle. The transition between oesophagus and intestine is gradual. There are no peptonephridia, oesophageal appendages or intestinal diverticula. Two types of coelomocyte are present.

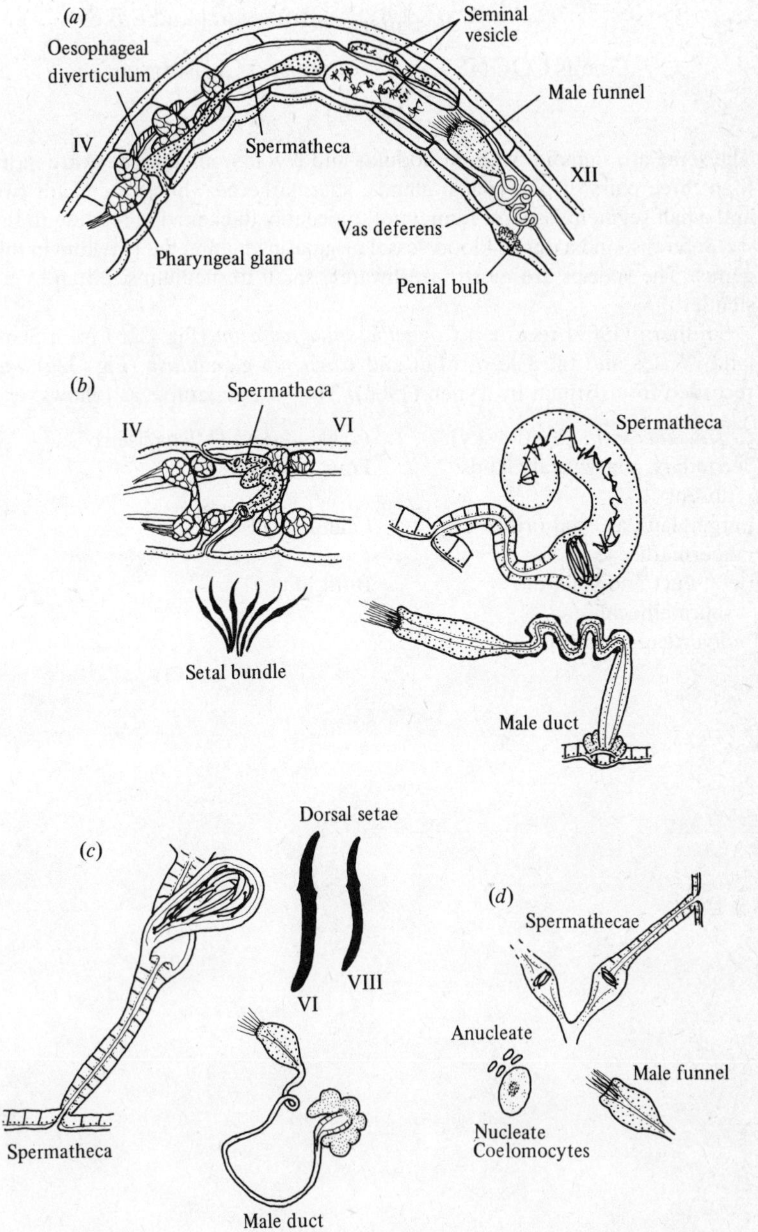

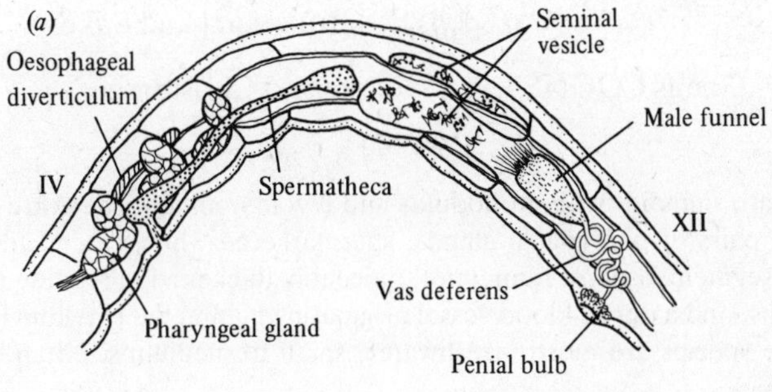

Fig. 31. The enchytraeids: *Achaeta littoralis* (*a*); *Cernosvitoviella immota* (*b*); *Mesenchytraeus armatus* (*c*); and *Hemifridericia parva* (*d*). (*a* after Lasserre, 1968, *b–d* after Nielsen and Christensen, 1959.)

B. Other genera

Genus COGNETTIA Nielsen and Christensen

(Fig. 32)

The setae are sigmoid, without nodulus and few in number. There are more than three pairs of pharyngeal glands, spermathecae which extend for two and a half segments, sperm funnel not noticeably thickened in relation to the vas deferens, and a dorsal blood vessel originating behind the clitellum in this genus. The species are mostly freshwater, small to medium-sized and very slender.

Southern (1909) recorded *Cognettia sphagnetorum* (Fig. 32*b*) from Scotland, Wales and the Isle of Man and *Cognettia glandulosa* (Fig. 32*a*) was recorded from Britain by Tynen (1966). They are separable as follows:

C. sphagnetorum (Vejdovsky)	*C. glandulosa* (Michaelsen)
Secondary pharyngeal glands absent	Four pairs present
Large gland at ectal orifice of spermathecae	Gland small
Ectal duct and stalk of spermathecal diverticulum short	Both long

(a)

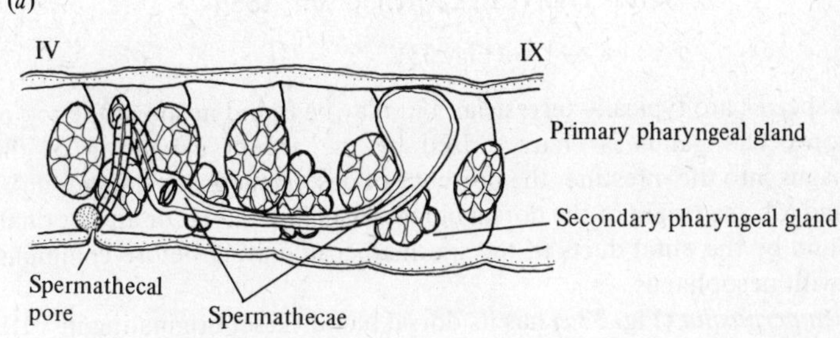

(b)

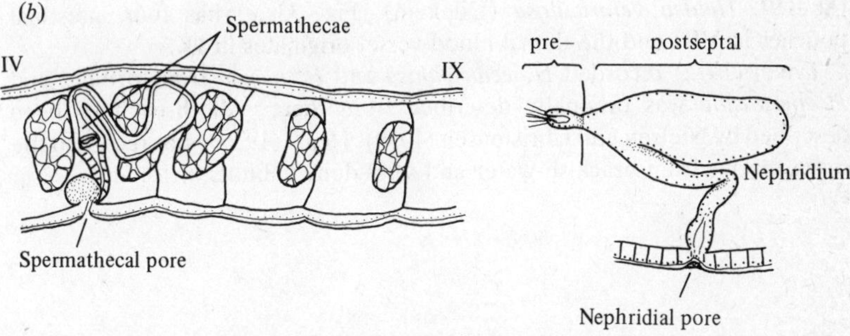

Fig. 32. Reproductive systems of (a) *Cognettia glandulosa* and (b) *C. sphagnetorum* (after Nielsen and Christensen, 1959).

Genus HENLEA Michaelsen, 1889

(Fig. 33)

Henlea species are typically terrestrial, but may be found in soil at the top of the shore. The genus is characterized by the abrupt expansion of the oesophagus into the intestine, the presence of oesophageal appendages in VI or IV and VI, the origin of the dorsal blood vessel in VIII–IX or anterior half of IX, and by the ental ducts of the spermathecae uniting before communicating with oesophagus.

Henlea perpusilla (Fig. 33*a*) has its dorsal blood vessel originating in VIII, as in *Henlea nasuta* (Eisen) (Fig. 33*b*,*d*), but the former has no intestinal diverticula between the oesophagus and intestine, whereas the latter has two (at 8/9). *Henlea ventriculosa* (Udekem) (Fig. 33*c*,*e*) has four intestinal pouches in VIII and the dorsal blood vessel originates in IX.

Tynen (1972) recorded *H. ventriculosa* and *H. nasuta* from Britain, and *H. perpusilla* was originally described from there. All three have been described by Nielsen and Christensen (1959). Healy (1979*b*) has recorded the three species from brackish-water and sand-dune habitats in Ireland.

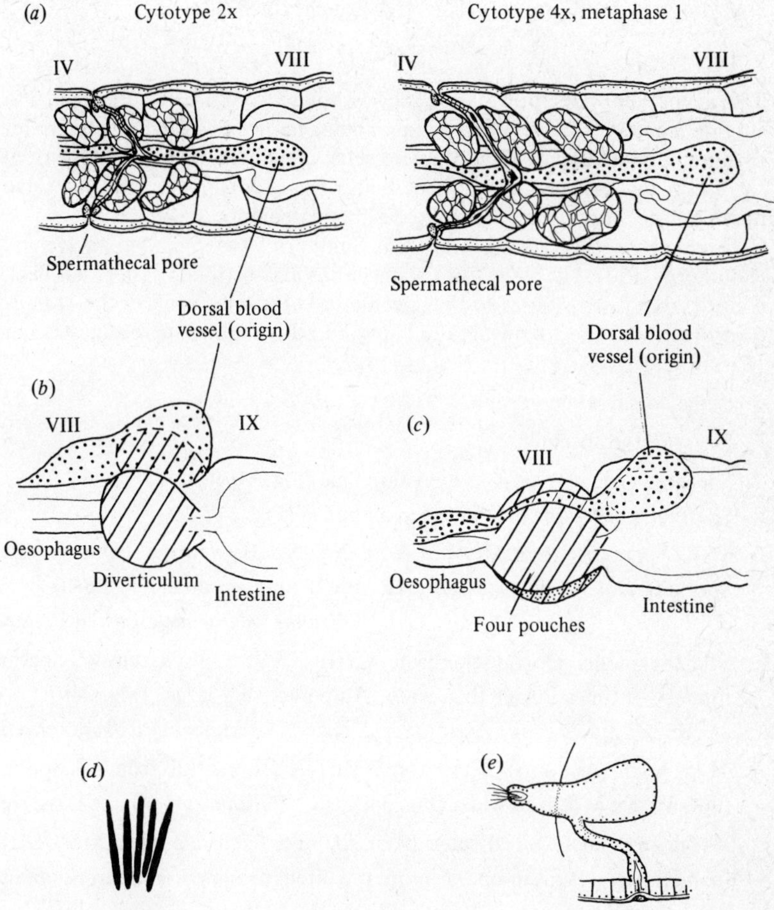

Fig. 33. *Henlea* spp.: (*a*) spermathecae of *H. perpusilla*; (*b*) anterior gut of *H. nasuta* and of (*c*) *H. ventriculosa*; (*d*) setae of *H. nasuta*; (*e*) nephridium of *H. ventriculosa* (after various authors).

Genus GRANIA Southern, 1913

(Fig. 34)

This genus comprises fully marine species, found from the lower intertidal to subtidal habitats. The setae are unique, being large and straight, with broad, recurved proximal ends (Fig. 5*l*). The setae are single, and are absent from the most anterior segments, there being more dorsal than ventral setae missing. The worms have a nematode-like appearance.

The type species, *Grania maricola* Southern, was recorded in Ireland (Southern, 1913), but other European species may be found. There has been considerable discussion as to the specific or subspecific rank of the various taxa described; the following key being based on the most recent account (Erseus and Lasserre, 1976; Erseus, 1977).

1. Some dorsal setae present **2**

Dorsal setae absent **5**

2. Spermathecae with pear-shaped ampullae (Fig. 34*b*) **3**

Spermathecae with constriction in ampullae (Fig. 34*d*) **4**

3. 37–49 segments. Dorsal setae from XX–XXIII, ventrals from V–VII. Sperm funnel 3–5 times longer than wide. Ampullae small. (Fig. 34*a*)

Grania macrochaeta pusilla Erseus

54–62 segments. Dorsal setae from XXIII–XXV, ventrals from VI. Sperm funnel 8–9 times longer than wide. Ampullae very large. (Fig. 34*b*)

. . . . *Grania maricola* Southern

4. 41–54 segments. Dorsal setae from XVIII–XXII, ventrals from IV. Sperm funnel twice as long as wide. (Fig. 34*d*) . . . . . *Grania roscoffensis* Lasserre

39–55 segments. Dorsal setae from XL or L, ventrals from XIV–XXII (sometimes present in one or more preclitellar segments). Sperm funnels several times longer than broad. (Fig. 34*e*)

Grania variochaeta Erseus and Lasserre

5. Ampullae of spermathecae sacciform. Ventral setae begin in XIII–XVII. (Fig. 34*f*,*g*) . . . . *Grania postclitellochaeta* (Knöllner)

Ampullae of spermathecae long, an egg-shaped oval. Ventral setae begin in XIII–XIV. (Fig. 34*c*,*h*) . . . . *Grania ovitheca* Erseus

The single British species, *G. maricola* is known from Clare Island, Ireland, at a depth of 40 m. It has been described, with most other potentially British species, by Erseus and Lasserre (1976), the other being described by Erseus (1977). The European species have been described from Scandinavia and France, and this indicates our lack of knowledge of the group as the two current authors are Swedish and French, respectively.

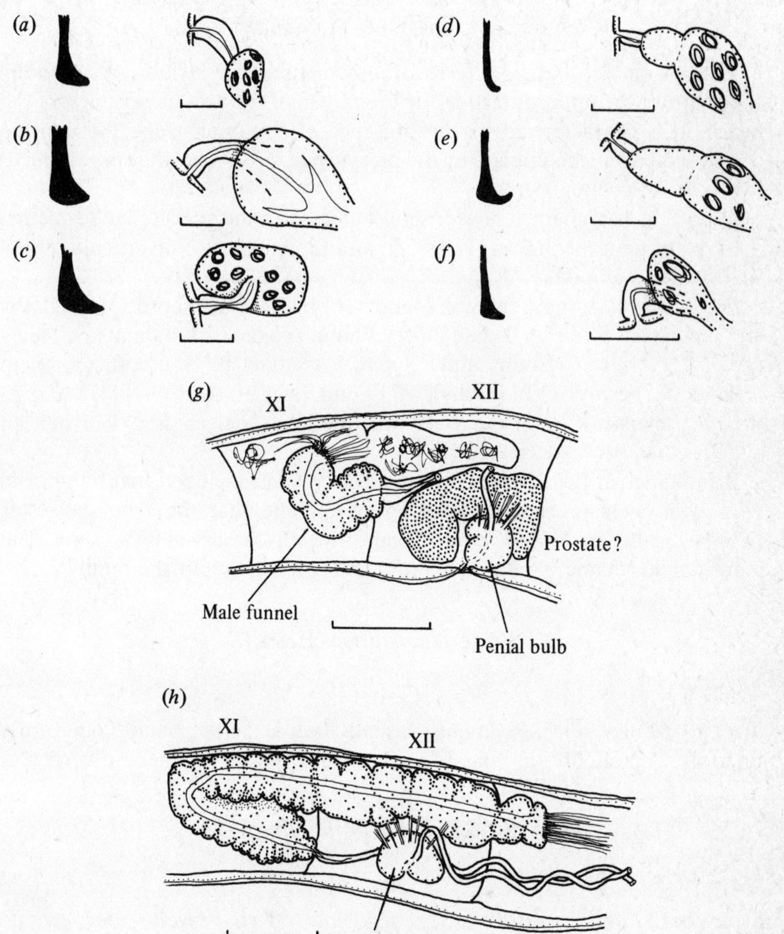

Fig. 34. *Grania* spp.: (*a–f*) setal base (left) and spermatheca (right) of (*a*) *G. macrochaeta pusilla*, (*b*) *G. maricola*, (*c*) *G. ovitheca*, (*d*) *G. roscoffensis*, (*e*) *G. variochaeta*, (*f*) *G. postclitellochaeta*; (*g*) reproductive system of *G. postclitellochaeta*; (*h*) reproductive system of *G. ovitheca* (after Erseus and Lasserre, 1976, and Erseus, 1977). Scale lines: (*a–f*), 50 μm; (*g*),(*h*), 100 μm.

Genus FRIDERICIA Michaelsen, 1889

This genus is also basically terrestrial, sometimes freshwater, occasionally supra-littoral or upper intertidal or found where freshwater runs across the beach. It is characterized by the presence of peptonephridia, the unusual paired setal arrangement and by the presence of dorsal coelomic pores and of two types of coelomocytes.

Important key characters are: setal numbers in the preclitellar segments, the type of peptonephridia (Fig. 35*f*), and the number of diverticula on the spermathecal ampullae.

Fridericia callosa (Eisen) was found very high on the shores of Anglesey and the Menai Strait by Tynen (1966) and in similar Irish habitats by Healy (1979*b*); *F. ratzeli* (Eisen) and *F. paroniana* Issel by Southern (1909) in Ireland; *F. perrieri* (Vejdovsky) by Friend (in Cernosvitov, 1941) and *F. striata* (Levinsen) in Wales – all described by Nielsen and Christensen (1959) – and other species may be found.

Brinkhurst (in Brinkhurst and Jamieson, 1971) concluded that by priority *Distichopus* has precedence over *Fridericia* and that the type species is *D. silvestris* Leidy. *Fridericia* was a name illegally conserved by Apstein. The more familiar name is retained here pending a revision of the family.

Fridericia callosa (Eisen)

(Fig. 35*a*)

Size, 10–20 mm, 46–58 segments. Usually four setae per bundle, anteriorly up to six. Peptonephridia coarse, 3–6 branches. No spermathecal diverticula.

Fridericia paroniana Issel

(Fig. 35*b*)

Size, 8–12 mm, (30)38–45(52) segments. Two, rarely four, setae. Peptonephridia mostly unbranched or 1–2 short terminal or subterminal branches. Two spermathecal diverticula, globular almost sessile.

Fridericia striata (Levinsen)

(Fig. 35*c*)

Size, 10–20 mm, (46)48–55(64) segments; 6–8(10) setae. Peptonephridia irregular, much branched. No spermathecal diverticula.

Fridericia perrieri (Vejdovsky)

(Fig. 35*d*)

Size, 10–25 mm, (33)42–50(64) segments; 4–8, usually 6, setae. Peptonephridia much branched (especially towards posterior end) but may have single branches anteriorly. Two almost cylindrical diverticula, slightly swollen at apices.

Fridericia ratzeli (Eisen)

(Fig. 35*e*)

Size, 20–35 mm, (38)51–60(70) segments; (4)5–8(9) setae. Peptonephridia irregular, multibranched. Spermathecal diverticula (5)6–8(10), difficult to count.

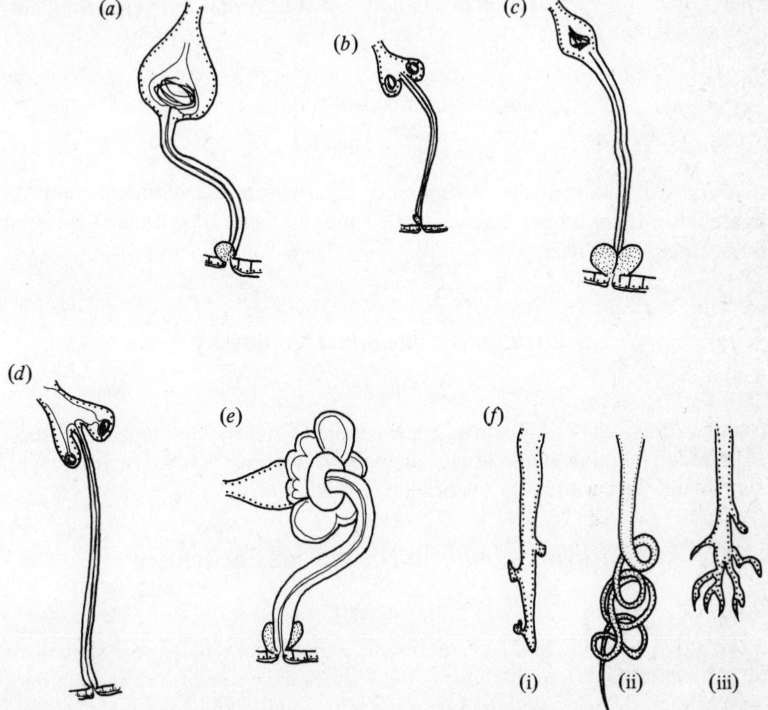

Fig. 35. Spermathecae of *Fridericia* spp.: (*a*) *F. callosa*; (*b*) *F. paroniana*; (*c*) *F. striata*; (*d*) *F. perrieri*; (*e*) *F. ratzeli*. Types of peptonephridia in *Fridericia* (*f*): (i) coarse, with short subterminal branches; (ii) tubular and with smooth surface, (iii) irregular and multibranched (after Nielsen and Christensen, 1959, and Tynen and Nurminen, 1969).

Genus ENCHYTRAEUS Henle, 1837

(Fig. 36)

A few *Enchytraeus* species are frequently found in intertidal habitats. There are four known British species, and a fifth possible entry, *Enchytraeus liefdeensis* Stephenson, recorded from Spitsbergen (Nurminen, 1965).

Enchytraeus characteristically have paired, unbranched peptonephridia and no dorsal coelomic pores apart from the head pore at 0/1. The setae are straight, with no nodulus.

Enchytraeus albidus Henle

(Fig. 36*a*)

Medium to large worms, 46–65 segments. Sperm funnel cylindrical, 5–8 times longer than wide, longer than diameter of XI. Penial bulb 'multi-lobed'. (Very common.)

Enchytraeus capitatus Bulow

(Fig. 36*b*)

Medium-sized worms, 50–54 segments. Sperm funnel cylindrical, approximately five times longer than wide, 1.5 times diameter of worm at XI. Penial bulb 'single', compact.

Enchytraeus buchholzi Vejdovsky

(Fig. 36*c*)

Small worms, 24–40 segments. Sperm funnel small, pear-shaped or almost cylindrical, length about 0.3 times diameter of worm at XI. Spermatheca with short ectal duct with a cluster of glands around oriface.

Enchytraeus minutus Nielsen and Christensen

(Fig. 36*d*)

Very small worms, 26–27 segments. Sperm funnel small, pear-shaped or almost cylindrical, length about 0.3–0.4 times diameter of XI. Spermatheca with a long, distinct ectal duct covered with small glandular cells.

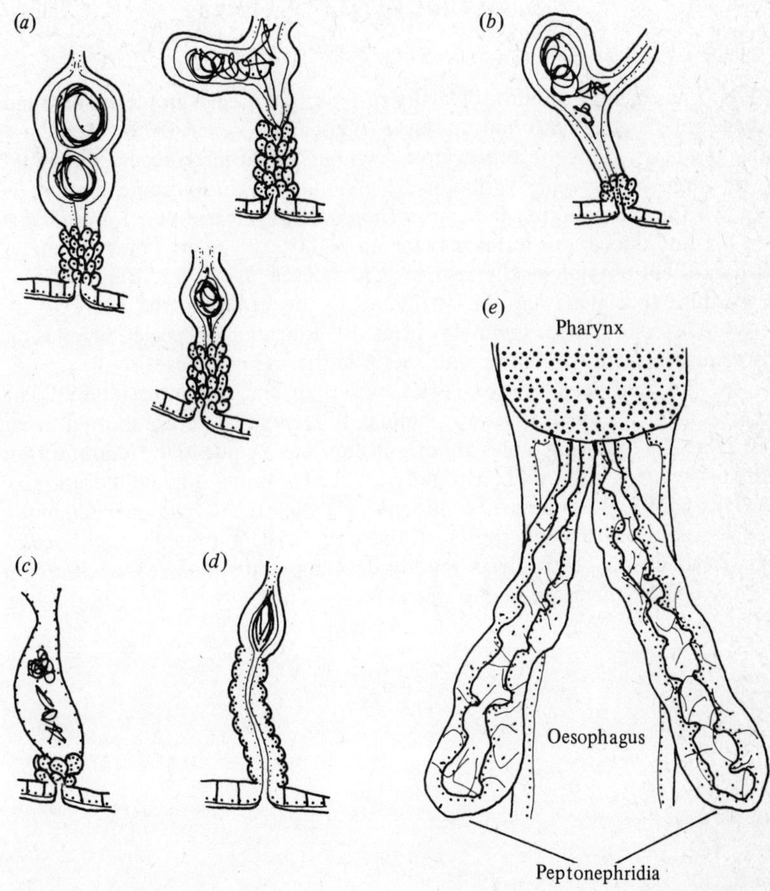

Fig. 36. *Enchytraeus* spp.: spermathecae of (*a*) *E. albidus*; (*b*) *E. capitatus*; (*c*) *E. buchholzi*; (*d*) *E. minutus*. Paired unbranched peptonephridia (*e*). (After various authors.)

Enchytraeus albidus is a cosmopolitan species, probably widely distributed in Britain, but recorded in the literature only from the Clyde and Ireland (McGrath, 1975); the species is common in decaying seaweed. *E. buchholzi* is widely distributed in Europe and is also reported from Ireland; it may prove to be cosmopolitan. *E. minutus* was recorded from Anglesey (Tynen, 1966, 1972) and is also quite widely distributed. The most recent descriptions can be found in Nielsen and Christensen (1959, 1961). Healy (1979*b*, *c*) reports *E. capitatus* from storm beaches and sand dunes in Ireland.

Genus LUMBRICILLUS Orsted

(Fig. 37)

This genus has the greatest diversity of intertidal species in the family, and they are frequently found in large numbers (Giere, 1975). They are taxonomically difficult, appearing to combine slight interspecific variability with large intraspecific variability. Variability in chromosome number is known for the pollution indicators *Lumbricillus lineatus* and *Lumbricillus rivalis*, and the various forms may be morphologically nearly, or completely, identical but physiologically distinct (Christensen, Jelnes and Berg, 1978).

Mature specimens may be attributed to the genus on the basis of the well-developed, lobed, seminal vesicles. Identification of species depends on obtaining mature specimens with sperm in the spermathecae.

All the European species keyed by Tynen and Nurminen (1969), bar *Lumbricillus fennicus* (Norway, Finland, Denmark), are reported from Britain. The following tables are meant simply as a guide to the identification of species, which should be carefully checked with the original literature.

Healy (1979*c*) has found *Lumbricillus semifuscus* to be the most common enchytraeid on exposed shores of the east coast of Ireland, *Lumbricillus kaloensis* characteristically in muddy deposits, and *Lumbricillus viridis* on sandy beaches near the high-water mark.

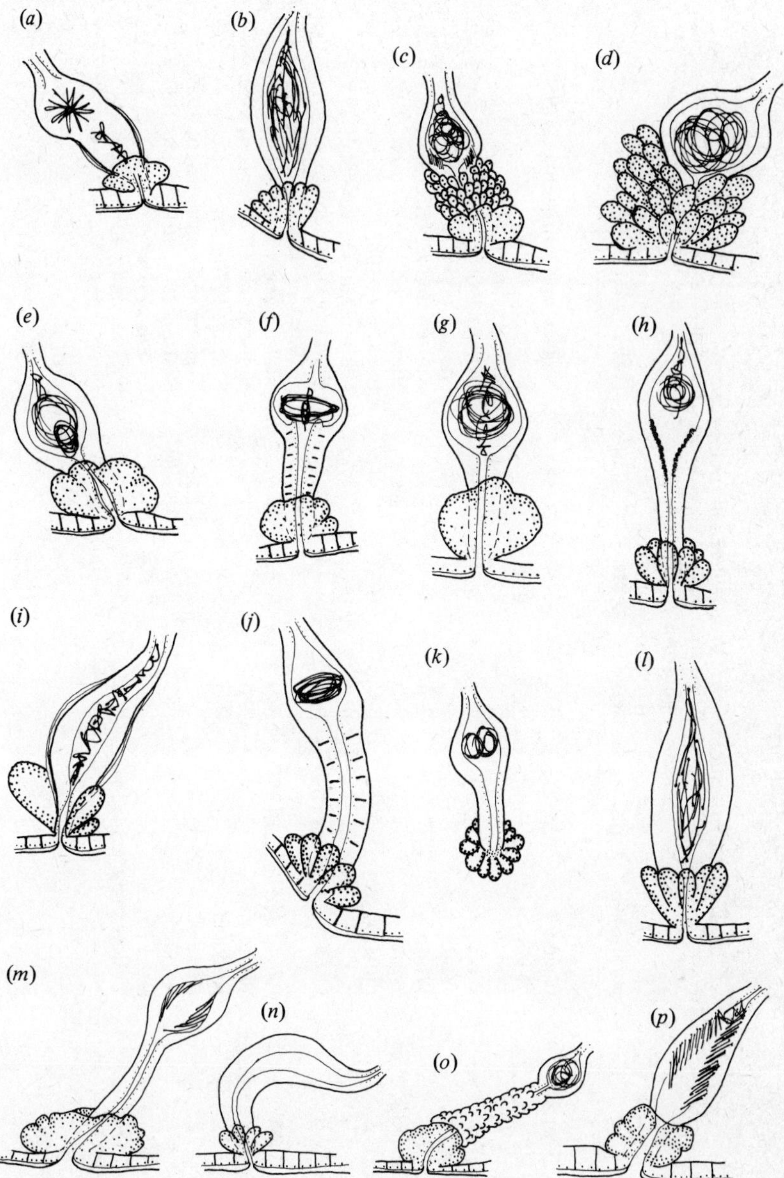

Fig. 37. Spermathecae of *Lumbricillus* spp. (after various authors): (*a*) *L. lineatus*; (*b*) *L. rivalis*; (*c*) *L. pagenstecheri*; (*d*) *L. reynoldsoni*; (*e*) *L. viridis*; (*f*) *L. tuba*; (*g*) *L. helgolandicus*; (*h*) *L. arenarius*; (*i*) *L. kaloensis*; (*j*) *L. knollneri* and *L. bulowi*; (*k*) *L. christenseni*; (*l*) *L. scoticus*; (*m*) *L. niger*; (*n*) *L. dubius* (apparently described from incompletely mature specimens); (*o*) *L. semifuscus*; (*p*) *L. pumilio*.

Table 2. *Characteristics of British marine* Lumbricillus *species*

Species	Colour (in life)	Size (mm)	Anterior dorsal setae	Anterior ventral setae	Setal shape*	Description
Lumbricillus						
viridis (Fig. 37*e*)	Green	20–25	3	5–9	ss	Nielsen and Christensen, 1959
niger (Fig. 37*m*)	Black	10–15	4–6	5–7	s	Southern, 1909
dubius (Fig. 37*n*)	White	12	2	2	st	Stephenson, 1911
christenseni (Fig. 37*k*)	White	5–9	2	2	st	Tynen, 1966
knollneri (Fig. 37*j*)	White	6–8	2	2	st	Nielsen and Christensen, 1959
tuba (Fig. 37*f*)	White	12	2–4	3–5	ss	Nielsen and Christensen, 1959
helgolandicus (Fig. 37*g*)	White	12–15	3–6	5–7	s	Nielsen and Christensen, 1959
reynoldsoni (Fig. 37*d*)	Pink or grey	25–30	5–7	6–9	ss	Backlund, 1948
lineatus (Fig. 37*a*)	Red	10–15	4–6	6–8	s	Nielsen and Christensen, 1959
scoticus (Fig. 37*l*)	Orange	7–9	6–8	10–14	s	Elmhirst and Stephenson, 1926 (see Erseus, 1977)
rivalis (Fig. 37*b*)	Bright red	20–35	6–9	7–12	s	Nielsen and Christensen, 1959
kaloensis (Fig. 37*i*)	White/red	12	3–6	4–7	s	Nielsen and Christensen, 1959
pumilio (Fig. 37*p*)	Reddish	2–3	4–6	5–8	s	Erseus, 1976*a*
arenarius (Fig. 37*h*)	Reddish	15	2–3	2–3	ss	Nielsen and Christensen, 1959
bulowi (Fig. 37*j*)	Reddish	12	2–3	2–3	ss/st	Nielsen and Christensen, 1959
semifuscus (Fig. 37*o*)	Bright red	16	2–7+	2–7+	s	Erseus, 1976*a*
pagenstecheri (Fig. 37*c*)	Yellow/brown	14–18	4–6	5–7	ss	Nielsen and Christensen, 1959

* s = sigmoid; ss = semi-sigmoid; st = straight.

Table 3. *British records of* Lumbricillus *species*

Species	Localities	References
Lumbricillus		
viridis	Wemyss Bay (Scotland), Anglesey, Dublin, Wexford	Stephenson, 1911; Tynen, 1972; McGrath, 1975; Healy, 1979*b*, *c*
niger	Dublin, Anglesey	Southern, 1909; Tynen, 1972
dubius	Wemyss Bay	Stephenson, 1911
christenseni	Anglesey	Tynen, 1966, 1972
knollneri	Anglesey	Tynen, 1966, 1972
tuba	Millport (Scotland), Anglesey, Gwynedd	Stephenson, 1911; Tynen, 1972
helgolandicus	Anglesey, Gwynedd	Tynen, 1972
reynoldsoni	Anglesey, Gwynedd	Tynen, 1972
lineatus	Dublin, Anglesey, Gwynedd, Scotland	Southern, 1909; Tynen, 1972; McGrath, 1975; Stephenson, 1922; Healy, 1979*b*
scoticus	Plymouth, Dublin, Anglesey, Cumbraes (Scotland)	Stephenson, 1932; Tynen, 1972; McGrath, 1975; Elmhirst and Stephenson, 1926
rivalis	Ireland, Scotland, Wales, Isle of Man	Southern, 1909; Stephenson, 1912; Tynen, 1972; Healy, 1979*b*, *c*
kaloensis	Anglesey, Gwynedd, Dublin	Tynen, 1966, 1972; McGrath, 1975; Healy, 1979*b*
pumilio	Plymouth	Stephenson, 1932
arenarius	Anglesey, Galway	Tynen, 1966, 1972; Healy, 1979*b*
bulowi	Anglesey, Wexford	Tynen, 1966, 1972; Healy 1979*c*
semifuscus	Fintry Bay (Scotland), Dublin, Wexford	Stephenson, 1912; Southern, 1909; Healy, 1979*c*
pagenstecheri	Baldoyle (Ireland), Lancashire, Anglesey, Gwynedd, Galway	Southern, 1909; Tynen, 1972; Healy 1979*b*

Genus MARIONINA Michaelsen

(Fig. 38)

There are a number of littoral and some sublittoral marine species in this genus. Four of the British species (*Marionina sjaelandica, M. southerni, M. spicula, M. subterranea*) are littoral and intertidal, and were reported from North Wales by Tynen (1972) (*M. southerni* was recorded from the Clyde as *Enchytraeus lobatus* by Southern, 1909). Three other marsh or lake-dwelling forms may be found on the upper shore, being reported from North Wales by Tynen (1972) (*M. appendiculata, M. argentea, M. communis*).

Lasserre (1964) described three new marine forms from France, and one (*M. achaeta*) which has been found on four continents, possibly because its lack of setae makes it instantly recognizable (apart from the genus *Achaeta* and the recently described asetate *Marionina arenaria* (Fig. 38*h*) from Ireland (Healy, 1979*a*)). *M. arenaria* Healy is a larger species than *M. achaeta* (Hagen), recently recorded from Ireland, and the spermathecae are not attached to the gut, as they are in *M. achaeta.* Erseus (1976*b*) has also described *Marionina sublitoralis* from Scandinavia, but these latter six are not keyed here. Healy (1979*b*, *c*) records that *M. preclitellochaeta* is common on southeastern Irish beaches (and probably also in Britain generally) and that *M. appendiculata* is common on exposed rocky shores.

1. Worms intensely white in reflected light (due to coelomocytes). Dorsal setae absent from II **2**

Worms not intensely white, dorsal setae all absent or all present **3**

2. Seminal vesicles present. Spermathecae as in Fig. 38*c* with diverticula

Marionina southerni (Cernosvitov)

Seminal vesicles absent. Spermathecae without diverticula (Fig. 38*d*)

Marionina argentea (Michaelsen)

3. Setae, where present, two (rarely three) per bundle **4**

Setae more than two in some bundles **6**

4. All dorsal setae absent **5**

Dorsal setae present

Marionina sjaelandica Nielsen and Christensen (Fig. 38*b*)

5. Ventral setae in II–VI only

Marionina preclitellochaeta Nielsen and Christensen (Fig. 38*g*)

All ventral setae present . . . . *Marionina subterranea* (Knöllner) (Fig. 38*f*)

6. Setae sigmoid (spermathecae as in Fig. 38*e*)

Marionina appendiculata Nielsen (Fig. 38*e*)

Setae straight **7**

7. Spermathecae communicating separately with oesophagus (as in Fig. 38*a*). Vas deferens wide, only 1.5–2 times the length of the sperm funnel
Marionina spicula (Leuckart) (Fig. 38*a*)

Spermathecae with ental ducts united at point of communication with oesophagus (as in Fig. 38*i*). Vas deferens narrow
Marionina communis Nielsen and Christensen (Fig. 38*i*)

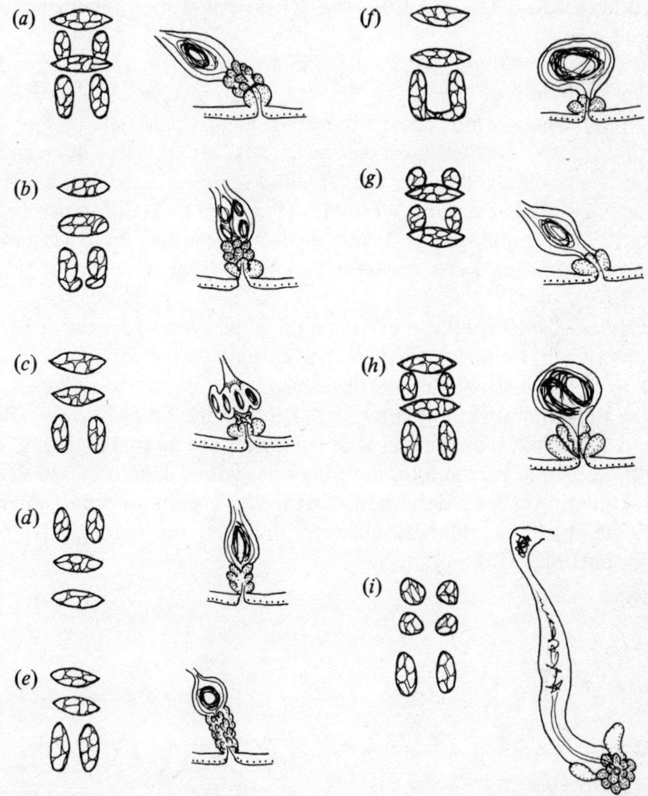

Fig. 38. Pharyngeal glands (left) and spermathecae (right) of *Marionina* spp. (after various authors): (*a*) *M. spicula*; (*b*) *M. sjaelandica*; (*c*) *M. southerni*; (*d*) *M. argentea*; (*e*) *M. appendiculata*; (*f*) *M. subterranea*; (*g*) *M. preclitellochaeta*; (*h*) *M. arenaria*; (*i*) *M. communis*.

Family AEOLOSOMATIDAE

(Fig. 39)

Although I do not regard this family as belonging in the Oligochaeta, because these miniature worms essentially share only their segmented, coelomate, setate condition with other annelids and have no other significant features in common with the true oligochaetes, the family (or families if the genus *Potamodrilus* be a separate family) is included here for convenience as it would otherwise be omitted from the series, being thus far unclaimed by anyone else.

Aeolosoma maritimum Westheide and Bunke, 1970, found on a sandy beach in the Gulf of Tunis, is the only marine species (salinity range 29.2–34.6‰). The worm is up to 4 mm long, has a single field of cilia on the ventral side of the prostomium and none of the ciliary pits that are sometimes found; it has orange-yellow epidermal glands. The setae are all hair-like in the dorsal as well as the ventral bundles (the main setal difference between these and all other oligochaetes), and are 3–5 per bundle, up to 115 μm long. The intestine widens between setal bundles 2–3 or 3–4 (Westheide and Bunke, 1970).

Aeolosoma litorale Bunke, 1967, is found in the Weser River near Bremen, Germany (where other salt-water species have been found, presumably related to the industrial pollution present), and also in some small rivers along the Baltic coast (salinity up to 5 ‰) and on the Finnish coast. The first zooid is 1.6–1.8 mm long, chains of seven zooids are up to 4 mm long. There are 7–10 hair setae per bundle, the long ones with a length of 110–150 μm. Sexual forms have been described, having three pairs of spermathecae in II–IV. The intestine enlarges between the posterior part of III to the posterior part of VII.

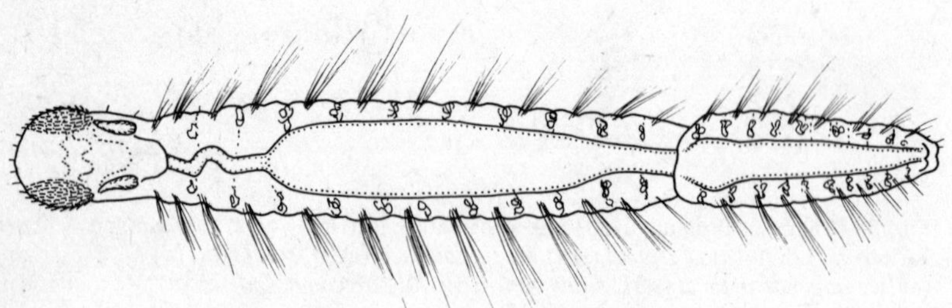

Fig. 39. *Aeolosoma*, whole animal.

Glossary

Ampulla Dilated region of, e.g., a spermatheca.

Anteseptale That part of a nephridium in front of the septum supporting the funnel.

Atrium The organ into which the vas deferens delivers sperm, also receives the prostate gland and leads to the male pore; function presumably nutrition and storage of sperm.

Bundles (of setae) Setae are arranged in four 'bundles' or groups typically on each segment except II, but setae may be missing from some – or even all – segments.

Clitellum Saddle-shaped epidermal gland, one cell thick, in the region of the genital pores in aquatic species; it secretes the cocoon.

Cocoon Egg case secreted by the clitellum; occurs loose in the sediments and contains several yolky fertilized eggs.

Coelom Fluid-filled body cavity lined by peritoneum.

Coelomocytes Large granular macronucleate cells in the coelomic fluid.

Copulatory bursa An involution of the ventral body wall around the male pore(s); often unpaired and median.

Cuticle External (secreted) layer of the body wall.

Dorsal coelomic pore Sphinctered connection between each coelomic compartment and the exterior, often situated mid-dorsally in the intersegmental groove. Mainly found in essentially terrestrial species, e.g. in the enchytraeid *Fridericia* in which they occur in all segments posteriorly from VII (rarely from VI).

Ectal Outermost; as applied to atria, that part nearest the penes. cf. **ental**.

Ejaculatory duct Thin extension of the atrium linking it to the penis or male pore.

Ental Innermost; as applied to atria, that part nearest the vasa deferentia and furthest from the male pore. cf. **ectal**.

Gonads Testes and ovaries; fundamentally two pairs of each in successive segments but reduced to one pair of testes and one pair of ovaries in adjacent segments in most aquatic oligochaetes.

Gut diverticula Simple extensions of the gut as blind pouches, especially in *Limnodriloides*.

Hair setae Slender elongate hair-like setae restricted to the dorsal bundles when present.

Head pore Communication between the coelom and the exterior situated on the prostomium at or near the anterior end, or at the transition between pro- and peristomium; may be absent in some species.

Male funnel Ciliated opening into male reproductive duct; leads to vas deferens; supported on appropriate septum, e.g. usually 10/11 in tubificids.

Male pore Superficial opening of male reproductive system; usually on XI in tubificids, usually ventrolateral in position where not fused as a median pore in a few species.

Nephridia (sing. **Nephridium**) Excretory tubes, usually with a ciliated funnel opening in one segment and the excretory pore in the segment behind.

Nodulus Swelling on seta, especially where it emerges from the setal sac to project beyond body wall.

Oesophagus Region of gut between pharynx and intestine.

Oesophageal appendage Sac-like evagination of the oesophagus in some enchytraeids.

Paratomy Reproduction by complex fission forming chains of individuals not in sequence according to age.

Penis Permanent fold of body wall enclosing the terminal section of the atrium/ejaculatory duct; often enclosed in specially thickened cuticular sheath.

Peptonephridia Structures found in association with the anterior part of the gut, doubtfully related to nephridia; sometimes termed salivary glands or oesophageal peptonephridia in enchytraeids.

Peristomium The first (seta-less) segment behind the pre-segmental prostomium.

Pharyngeal gland Aggregation of the cell bodies of cells from the lining of the eversible pharyngeal roof which extend back to lie around the gut in IV, V, VI, etc.

Postseptale That part of the nephridium behind the septum.

Proboscis Special elongation of the prostomium.

Prostate gland Cell bodies of glandular cells originally lining the atrium which extend through the muscular atrial wall to lie in the coelom.

Prostomium Terminal anterior portion of annelid; pre-segmental.

Pseudopenis Temporary penial device, either protrusible (usually the distal end of the atrium) or eversible.

Segment A unit of the body, separated from other such units by septa, each unit (potentially) with its own setae, nephridia, lateral blood vessels, nerves, etc.; functionally supposed to permit continuous burrowing. Superficial annulation not always strictly related to the internal segmentation in that additional annulae may occur.

Seminal vesicle See **sperm sac.**

Septum (pl. **Septa**) Muscular sheet separating two segments, functions as a fluid-resistant bulkhead.

Setae (sing. **Seta**) Bristles, each the product of a single secretory cell, used as aids to locomotion and (rarely) other functions such as reproduction and feeding. Setae emerge from setal sacs. Also termed chaetae.

Sperm funnel See **male funnel.**

Sperm sac Pocket on the anterior and posterior septum of the testicular segment in which sperm undergoing the later stages of spermatogenesis are stored; when particularly large usually termed seminal vesicle.

Spermatheca Sperm-receiving and storage device filled by concopulant during mating; presumably provides sperm for fertilizing eggs in the cocoon.

Spermatophore Sperm-containing structure attached to the body wall of a concopulant in species lacking spermathecae.

Spermatozeugmata Complex sperm bundles with the sperm heads internal, the tails external and arranged in a spiral, all held together in a matrix. During copulation placed in spermathecae.

Vas deferens (pl. **Vasa deferentia**) Sperm-carrying canal extending from the male funnel to the atrium.

Acknowledgements

I am deeply indebted to Ms K. Coates, who provided voluminous notes from which I was able to construct the section of this manuscript dealing with the Enchytraeidae. Many British colleagues shared their knowledge of worm distributions, some of it unpublished, which was especially valuable as this work was completed in Canada. Dr C. Erseus generously shared his knowledge, especially of the phallodriline tubificids, a group he is in process of revising. Dr B. Healy kindly corrected and amended parts of the enchytraeid section.

Miss A. Sinclair helped with the illustrations, Mrs S. McKenzie and Mrs N. Collier typed the manuscript. The laboratory of Analytical Systematics of the Royal Ontario Museum, Toronto, sectioned many specimens for me.

References

Aarefjord, F., Borgstrom, R., Lien, L. and Milbrink, G. (1973). Oligochaetes in the bottom fauna and stomach content of trout (*Salmo trutta* (L.)). *Norwegian J. Zool.*, **21,** 281–8.

Arlt, G. (1969). Bemerkungen zur Bodenfauna des Greifswalder Boddens. *Wiss. Zeit. E. M. Arndt-Univ. Greifswald*, **18,** 189–93.

Aston, R. J. (1968). The effect of temperature on the life cycle growth, and fecundity of *Branchiura sowerbyi* (Oligochaeta, Tubificidae). *J. Zool.*, **154,** 29–40.

Aston, R. J. (1973). Tubificids and water quality: a review. *Environ. Pollut.*, **5,** 1–10.

Backlund, H. O. (1948). *Lumbricillus reynoldsoni* n. sp., an enchytraeid worm from the beaches of North Wales. *J. mar. biol. Ass. U.K.*, **27,** 710–17.

Bagge, P. (1969). Effects of pollution on estuarine ecosystems. *Merentutkimuslaitoksen Julkaisu*, **228,** 3–118.

Baker, H. R. (1980). A redescription of *Tubificoides pseudogaster* (Dahl) (Oligochaeta, Tubificidae). *Trans. Am. microsc. Soc.*, **99,** 337–42.

Baker, H. R. and Erseus, C. (1979). *Peosidrilus biprostatus* n.g., n.sp. A marine tubificid (Oligochaeta) from eastern United States. *Proc. biol. Soc. Wash.*, **92,** 505–9.

Baker, H. R. and Brinkhurst, R. O. (1981). A revision of the genus *Monopylephorus* and redefinition of the subfamilies Rhyacodrilinae and Branchiurinae (Tubificidae, Oligochaeta). *Can. J. Zool.*, **59,** 939–65.

Birtwell, I. and Arthur, D. R. (1980). The ecology of the tubificids in the Thames estuary, with particular reference to *Tubifex costatus* (Claparede). In *Aquatic oligochaete biology*, ed. R. O. Brinkhurst and D. G. Cook, pp. 331–81 Plenum Press, New York.

Bonomi, G. and DiCola, G. (1980). Population dynamics of *Tubifex tubifex* studied by means of a new model. In *Aquatic oligochaete biology*, ed. R. O. Brinkhurst and D. G. Cook, pp. 185–204. Plenum Press, New York.

Brinkhurst, R. O. (1962). A check list of British Oligochaeta, *Proc. zool. Soc. Lond.*, **138,** 317–30.

Brinkhurst, R. O. (1963*a*). A genus of brackish water Oligochaeta new to Britain. *Nature*, 1206.

Brinkhurst, R. O. (1963*b*). Notes on the brackish water and marine species of Tubificidae (Annelida, Oligochaeta). *J. mar. biol. Ass. U.K.*, **43,** 709–15.

Brinkhurst, R. O. (1963*c*). A guide for the identification of British aquatic Oligochaeta. *Scient. Publs. Freshwat. biol. Ass.*, **22,** 1–52.

Brinkhurst, R. O. (1964). Observations on the biology of the marine Oligochaeta *Tubifex costatus*. *J. mar. biol. Ass. U.K.*, **44,** 11–16.

Brinkhurst, R. O. (1966*a*). The Tubificidae of polluted waters, *Verh. Internat. Verein. Limonol.*, **16,** 854–9.

Brinkhurst, R. O. (1966*b*). A contribution to the systematics of the marine Tubificidae (Annelida, Oligochaeta). *Biol. Bull.*, **130,** 297–303.

Brinkhurst, R. O. (1974). *The benthos of lakes*. MacMillan Press, London. 190 pp.

Brinkhurst, R. O. (1978). Freshwater Oligochaeta in Canada. *Can. J. Zool.*, **56,** 2166–75.

Brinkhurst, R. O. (1981). Oligochaeta [in part]. In *Taxonomy and classification of living organisms*, ed. S. P. Parker. McGraw Hill, New York.

Brinkhurst, R. O. and Baker, H. R. (1979). A review of the marine Tubificidae (Oligochaeta) of North America. *Can. J. Zool.*, **57,** 1553–69.

Brinkhurst, R. O. and Chuan K. E. (1969). Preliminary investigation of the exploitation of some potential nutritional resources by three sympatric tubificid oligochaetes. *J. Fish. Res. Bd Can.*, **26,** 2659–68.

Brinkhurst, R. O., Chuan K. E. and Kaushik, N. (1972). Interspecific interactions and selective feeding by tubificid oligochaetes. *Limnol. Oceanogr.*, **17,** 122–33.

Brinkhurst, R. O. and Cook, D. G. (eds.) (1980). *Aquatic oligochaete biology*. Plenum Press, New York.

Brinkhurst, R. O. and Jamieson, B. G. M. (1971). *Aquatic Oligochaeta of the world.* Oliver and Boyd, Edinburgh. 860 pp.

Bulow, T. (1955). Oligochaeten aus den Endgebieten der Schlei. *Kieler Meeresforsch.*, **11,** 253–64.

Bulow, T. (1957). Systematisch–autokologische Studien an eulitoralen Oligochaeten der Kimbrischen Halbinsel. *Kieler Meeresforsch.*, **13,** 69–116.

Bunke, D. (1967). Zur Morphologie und Systematik der Aeolosomatidae Beddard 1895 und Potamodrilidae nov. fam. *(Oligochaeta). Zool. Jb.*, **94,** 187–368.

Cain, A. J. (1959). Oligochaeta new to Britain. *Ann. Mag. nat. Hist.*, **2,** 193.

Cernosvitov, L. (1941). Revision of Friend's types and descriptions of British Oligochaeta. *Proc. zool. Soc. Lond.* (*b*), **111,** 237–80.

Chapman, P. M. and Brinkhurst, R. O. (1981). Seasonal changes in interstitial salinities and seasonal movements of subtidal benthic invertebrates in the Fraser River estuary, British Columbia. *Est. coastal mar. Sci.*, **12,** 49–66.

Chapman, P. M., Churchland, L. M., Thomson, P. A. and Michnowsky, E. (1980). Heavy metal studies with oligochaetes. In *Aquatic oligochaete biology*, ed. R. O. Brinkhurst and D. G. Cook, pp. 477–502. Plenum Press, New York.

Christensen, B., Jelnes, J. and Berg, U. (1978). Long-term isozyme variation in parthenogenetic forms of *Lumbricillus lineatus* (s.1.) (Enchytraeidae, Oligochaeta). *Hereditas*, **88,** 65–73.

Cook, D. G. (1969*a*). Observations on the life history and ecology of some Lumbriculidae (Annelida, Oligochaeta). *Hydrobiologia*, **34**, 561–74.

Cook, D. G. (1969*b*). The Tubificidae (Annelida, Oligochaeta) of Cape Cod Bay with a taxonomic revision of the genera *Phallodrilus* Pierantoni, 1902, *Limnodriloides* Pierantoni, 1903, and *Spiridion* Knöllner, 1935. *Biol. Bull.*, **136,** 9–27.

Cook, D. G. (1970*a*). Bathyal and abyssal Tubificidae (Annelida, Oligochaeta) from the Gay Head Bermuda transect, with descriptions of new genera and species. *Deep-Sea Res.*, **17,** 973–81.

Cook, D. G. (1970*b*). *Torodrilus lowryi*: new genus and species of marine tubificid Oligochaeta from Antarctica. *Trans. Am. microsc. Soc.*, **89,** 282–8.

Cook, D. G. (1970*c*). *Peloscolex dukei* n. sp. and *P. aculeatus* n. sp. (Oligochaeta, Tubificidae) from the North-West Atlantic, the latter being from abyssal depths. *Trans. Am. microsc. Soc.*, **88,** 492–7.

Cook, D. G. (1971). The Tubificidae (Annelida, Oligochaeta) of Cape Cod Bay II. Ecology and systematics, with the description of *Phallodrilus parviatriatus* nov. sp. *Biol. Bull.*, **141,** 203–21.

Cook, D. G. (1974). The systematics and distribution of marine Tubificidae

(Annelida, Oligochaeta) in the Bahia De San Quintin, Baja California, with descriptions of five new species. *Bull. South. Calif. Acad. Sci.*, **73,** 126–40.

Cook, D. G. and Hiltunen, J. K. (1975). *Phallodrilus hallae*, a new tubificid oligochaete from the St Lawrence Great Lakes. *Can. J. Zool.*, **53,** 934–41.

Cross, W. H. (1976). A study of predation rates of leeches on tubificid worms under laboratory conditions. *Ohio J. Sci.*, **76,** 164–6.

Dash, M. C. (1970). A taxonomic study of Enchytraeidae (Oligochaeta) from Rocky Mountain forest soils of the Kananaskis region of Alberta, Canada. *Can. J. Zool.*, **48,** 1429–35.

Dash, M. C. and Cragg, J. B. (1972). Selection of microfungi by Enchytraeidae (Oligochaeta) and other members of the soil fauna. *Pedobiologia*, **12,** 282–6.

Davis, R. B. (1974*a*). Tubificids alter profiles of redox potential and pH in profundal lake sediment. *Limnol. Oceanogr.*, **19,** 342–6.

Davis, R. B. (1974*b*). Stratigraphic effects of tubificids in profundal lake sediments. *Limnol. Oceanogr.*, **19,** 466–88.

Davis, R. B., Thurlow, D. L. and Brewster, F. E. (1975). Effects of burrowing tubificid worms on the exchange of phosphorous between lake sediment and overlying water. *Verh. Internat. Verein. Limnol.*, **19,** 382–94.

Dzwillo, M. (1966). Oligochaeten in Marinen Raum. *Veröff. Inst. Meeresforsch. Bremerh.*, **2,** 155–60.

Elmhirst, R. and Stephenson, J. (1926). On *Lumbricillus scoticus*. *J. mar. biol. Ass. U.K.*, **14,** 469–73.

Erseus, C. (1975). On the systematic position of *Rhyacodrilus prostatus* Knöllner (Oligochaeta, Tubificidae). *Zool. Scripta*, **4,** 33–5.

Erseus, C. (1976*a*). Littoral Oligochaeta (Annelida) from Eyjafjordur, north coast of Iceland. *Zool. Scripta*, **5,** 5–11.

Erseus, C. (1976*b*). Marine subtidal Tubificidae and Enchytraeidae (Oligochaeta) of the Bergen area, western Norway. *Sarsia,* **62,** 25–48.

Erseus, C. (1977). Marine Oligochaeta from the Koster area, west coast of Sweden, with descriptions of two new enchytraeid species. *Zool. Scripta*, **6,** 293–8.

Erseus, C. (1978*a*). Two new species of the little-known genus *Bacescuella* Hrabe (Oligochaeta, Tubificidae) from the North Atlantic. *Zool. Scripta,* **7,** 263–7.

Erseus, C. (1978*b*). New species of *Adelodrilus* and a revision of the genera *Adelodrilus* and *Adelodriloides* (Oligochaeta, Tubificidae). *Sarsia*, **63,** 135–44.

Erseus, C. (1979*a*). Re-examination of the marine genus *Spiridion* Knöllner (Oligochaeta, Tubificidae). *Sarsia*, **64,** 183–7.

Erseus, C. (1979*b*). Taxonomic revision of the marine genera *Bathydrilus* Cook and *Macroseta* Erseus (Oligochaeta, Tubificidae) with descriptions of six new species and subspecies. *Zool. Scripta*, **8,** 139–51.

Erseus, C. (1979*c*). *Coralliodrilus leviatriatus* gen. et sp. n., a marine tubificid (Oligochaeta) from Bermuda. *Sarsia*, **64,** 179–82.

Erseus, C. (1979*d*). *Uniporodrilus granulothecus* n.g.n.sp., a marine tubificid (Oligochaeta) from eastern United States. *Trans. Am. microsc. Soc.*, **98,** 414–18.

Erseus, C. (1979*e*). *Bermudrilus peniatus* n.g.n.sp. (Oligochaeta, Tubificidae) and two new species of *Adelodrilus* from the North West Atlantic. *Trans. Am. microsc. Soc.*, **98,** 418–27.

Erseus, C. (1979*f*). Taxonomic revision of the maine genus *Phallodrilus* Pierantoni with descriptions of thirteen new species. *Zool. Scripta*, **8,** 187–208.

Erseus, C. (1979*g*). *Inanidrilus bulbosus* gen. et sp. n., a marine tubificid (Oligochaeta) from Florida, U.S.A. *Zool. Scripta*, **8,** 209–10.

Erseus, C. (1980*a*). Specific and generic criteria in marine Oligochaeta with special emphasis on Tubificidae. In *Aquatic oligochaete biology*, ed. R. O. Brinkhurst and D. G. Cook, pp. 9–24. Plenum Press, New York.

Erseus, C. (1980*b*). Redescriptions of *Phallodrilus parthenopaeus* Pierantoni and *P. obscurus* Cook (Oligochaeta, Tubinicidae). *Zool. Scripta*, **9,** 93–6.

Erseus, C. (1980*c*). Taxonomic studies on the marine genera *Aktedrilus* Knöllner and *Bacescuella* Hrabe (Oligochaeta, Tubificidae), with descriptions of seven new species. *Zool. Scripta*, **9,** 97–111.

Erseus, C. (1981). Taxonomic studies of Phallodrilinae (Oligochaeta, Tubificidae) from the Great Barrier Reef and the Comoro Islands with descriptions of ten new species and one new genus. *Zool. Scripta*, **10,** 15–31.

Erseus, C. and Lasserre, P. (1976). Taxonomic status and geographic variation of the marine enchytraeid genus *Grania* Southern (Oligochaeta). *Zool. Scripta*, **5,** 121–32.

Eyres, J. P., Williams, N. V. and Pugh-Thomas, M. (1978). Ecological studies on Oligochaeta inhabiting depositing substrata in the Irwell, a polluted English river. *Freshwat. Biol.*, **8,** 24–32.

Finogenova, N. P. (1976). New species of oligochaete worms of the family Tubificidae from the Caspian Sea. *Zhol. Zh.*, **55,** 1563–6.

Gage, J. (1974). Shallow-water zonation of sea-loch benthos and its relation to hydrographic and other physical features. *J. mar. biol. Ass. U.K.*, **54,** 223–49.

Giere, O. (1970). Untersuchungen zur Mikrozonierung und Okologie mariner Oligochaeten im Sylter Watt. *Veröff. Inst. Meeresforsch. Bremerh.*, **12,** 491–529.

Giere, O. (1975). Population structure, food relations and ecological role of marine oligochaetes with special reference to meiobenthic species. *Mar. Biol.*, **31,** 139–56.

Giere, O. (1980). Tolerance and preference reactions of marine Oligochaeta in relation to their distribution. In *Aquatic oligochaete biology*, ed. R. O. Brinkhurst and D. G. Cook, pp. 385–409. Plenum Press, New York.

Gray, J. S. (1976). The fauna of the polluted River Tees Estuary. *Est. coastal mar. Sci.*, **4,** 653–76.

Green, J. (1954). A note on the food of *Chaetogaster diaphanus*. *Ann. Mag. nat. Hist.*, **1,** 842–4.

Grigelis, A. (1980). Ecological studies of aquatic oligochaetes in the U.S.S.R. In *Aquatic oligochaete biology*, ed. R. O. Brinkhurst and D. G. Cook, pp. 225–40. Plenum Press, New York.

Harman, W. J. (1977). Three new species of Oligochaeta (Naididae) from the south eastern U.S. *Proc. biol. Soc. Wash.*, **90,** 483–90.

Harman, W. J. and Loden, M. (1978). A re-evaluation of the Opistocystidae (Oligochaeta) with descriptions of two new genera. *Proc. biol. Soc. Wash.*, **91,** 453–62.

Harrison, J. and Grant, P. (1976). *The Thames transformed.* Deutsch. London. 240 pp.

Healy, B. (1979*a*). Three new species of Enchytraeidae (Oligochaeta) from Ireland. *Zool. J. Linn. Soc.*, **67,** 87–95.

Healy, B. (1979*b*). Records of Enchytraeidae (Oligochaeta) in Ireland. *J. Life Sci. R. Dubl. Soc.*, **1,** 39–70.

Healy, B. (1979*c*). Marine fauna of County Wexford. 1 – Littoral and brackish water Oligochaeta. *Ir. Nat. J.*, **19,** 418–22.

Holme, N. A. and McIntyre, A. D. (eds.) (1971). *Methods for the study of marine benthos*. Blackwell, Oxford.

Holmquist, C. (1978). Revision of the genus *Peloscolex* (Oligochaeta, Tubificidae). 1. Morphological and anatomical scrutiny, with discussion on the generic level. *Zool. Scripta*, **7**, 187–200.

Hrabe, S. (1960). Oligochaeta limicola from the collection of Dr S. Husmann. *Spisy Přír. Fak. Univ. v Brně*, **415**, 245–77.

Hrabe, S. (1967). Two new species of the family Tubificidae from the Black Sea, with remarks about various species of the subfamily Tubificinae. *Spisy Přír. Fak. Univ. v Brně*, **485**, 331–56.

Hrabe, S. (1971*a*). On new marine Tubificidae of the Adriatic Sea. *Scripta Fac. Scient. nat. Ujep. Brun. Biol.* (3), **1**, 215–26.

Hrabe, S. (1971*b*). A note on the Oligochaeta of the Black Sea. *Věst. Sc. Čs. Spol. Zool.*, **35**, 32–4.

Hrabe, S. (1973). A contribution to the knowledge of marine Oligochaeta mainly from the Black Sea. *Trav. Mus. Hist. Nat. G. Antipa*, **13**, 27–38.

Hunter, J. and Arthur, D. R. (1978). Some aspects of the ecology of *Peloscolex benedeni* Udekem (Oligochaeta, Tubificidae) in the Thames estuary. *Est. coastal mar. Sci.*, **6**, 197–208.

Jamieson, B. G. M. (1977). Marine meiobenthic Oligochaeta from Heron and Wistari Reefs (Great Barrier Reef) of the genera *Clitellio, Limnodriloides*, and *Phallodrilus* (Tubificidae) and *Grania* (Enchytraeidae). *Zool. J. Linn. Soc.*, **61**, 329–49.

Kasprazak, K. (1972*a*). Variability of a *Chaetogaster diaphanus* (Gruithuisen), 1828 (Oligochaeta, Naididae) in different environments. *Zool. Polon.*, **22**, 43–51.

Kasprazak, K. (1972*b*). *Enchytraeus mariae*, a new species of Enchytraeidae, Oligochaeta, found in the national park of Great Poland. *B. Pol. Biol.*, **21**, 279–84.

Kendall, M. A. (1979). The stability of the deposit feeding community of a mud flat in the River Tees. *Est. coastal mar. Sci.*, **8**, 15–22.

Kennedy, C. R. (1966*a*). The life history of *Limnodrilus udekemianus* Clap. (Oligochaeta, Tubificidae). *Oikos*, **17**, 10–18.

Kennedy, C. R. (1966*b*). The life history of *Limnodrilus hoffmeisteri* Clap. (Oligochaeta, Tubificidae) and its adaptive significance. *Oikos*, **17**, 158–68.

Kennedy, C. R. (1969). Tubificid oligochaetes as food of dace *Leuciscus leuciscus* (L.). *J. Fish Biol.*, **1**, 11–15.

Knöllner, F. H. (1935). Okologische und systematische Untersuchungen über litorale und Marine Oligochaten der Kieler Bucht. *Zool. Jb. (Syst.)*, **66**, 425–512.

Ladle, M. (1971). The biology of Oligochaeta from Dorset Chalk streams. *Freshwat. Biol.*, **1**, 83–97.

Lasserre, P. (1964). Note sur quelques Oligochetes Enchytraeidae présents dans les plages du bassin d'Archachon. *P. V. Soc. Linn. Bordeaux*, **101**, 1–5.

Lasserre, P. (1968). Présence du genre *Achaeta* Vejdovsky, 1877 (Oligochaeta, Enchytraeidae) dans des plages sableuses marines. *Bull. Mus. natn. Hist. nat.*, **39**(1967), 979–83.

Lasserre, P. (1976). Metabolic activities of benthic microfauna and meiofauna. In *The benthic boundary layer*, ed. I. N. McCave, pp. 95–142. Plenum Press, New York.

Lastockin, D. A. (1937). New species of Oligochaeta Limicola in the European part of the U.S.S.R. *Dok. Akad. Nauk. SSSR*, **17**, 233–5.

Learner, M. A., Lochhead, G. and Hughes, B. C. (1978). A review of the biology of British Naididae (Oligochaeta) with emphasis on the environment. *Freshwat. Biol.*, **8**, 357–75.

Leppakowski, E. (1975). Assessment of degree of pollution on the basis of macrozoobenthos in marine and brackish-water environments. *Acta Acad. Aboensis* (*B*), **35**(2), 1–90.
Loden, M. S. (1974). Predation by chironomid (Diptera) larvae on oligochaetes. *Limnol. Oceanogr.*, **19,** 156–9.
McCall, O. L. and Fisher, J. B. (1980). Effects of tubificid oligochaetes on physical and chemical properties of Lake Erie sediments. In *Aquatic oligochaete biology*, ed. R. O. Brinkhurst and D. G. Cook, pp. 253–317. Plenum Press, New York.
McElhone, M. J. (1978). A population study of littoral dwelling Naididae (Oligochaeta) in a shallow mesotrophic lake in North Wales. *J. Anim. Ecol.*, **47,** 615–26.
McGrath, D. (1975). Notes on some Irish marine littoral and freshwater Oligochaeta (Annelida). *Irish Nat. J.*, **18,** 216–18.
McLusky, D. S., Teare, M. and Phizachlea, P. (1981). Effects of domestic and industrial pollution on distribution and abundance of aquatic oligochaetes in the Forth estuary. *Proceedings of the 14th European Marine Biology Symposium, Helgoland,* 1977. *Helgoländer wiss. Meeresunters*, **33,** 113–21.
Marcus, E. (1965). Naidomorpha aus brasilianischem Brackwasser. *Beitr. Neotrop. Fauna*, **4,** 61–83.
Michaelsen, W. (1908). Zur Kenntnis der Tubificiden. *Arch. Naturgesch.*, **74,** 129–62.
Michaelsen, W. (1926). Oligochaeten aus dem Ryck bei Greifswald und von benachbarten Meeresgebieten. *Mitt. Hamb. Zool. Mus. Inst.*, **42,** 21–9.
Moore, J. P. (1905). Some marine Oligochaeta of New England. *Proc. Acad. nat. Sci. Philadelphia*, **57,** 373–99.
Moore, J. W. (1978). Importance of algae in the diet of the oligochaetes *Lumbriculus variegatus* (Müller) and *Rhyacodrilus sodalis* Eisen. *Oecologia*, **35,** 357–63.
Nielsen, C. O. and Christensen, B. (1959). The Enchytraeidae. Critical revision and Taxonomy of European species. *Nat. Jutl.*, **8–9,** 1–160.
Nielsen, C. O. and Christensen, B. (1961). Studies on Enchytraeidae VII. Critical revision and taxonomy of European species. Supplement 1. *Nat. Jutl.*, **10**(1961), 1–19.
Nielsen, C. O. and Christensen, B. (1963). Studies on Enchytraeidae VII. Critical revision and taxonomy of European species. Supplement 2. *Nat. Jutl.*, **10**(1963), 1–23.
Nurminen, M. (1965). Enchytraeid and lumbricid records (Oligochaeta) of Spitsbergen. *Ann. zool. fenn.*, **3,** 68–9.
Nurminen, M. (1970). Records of Enchytraeidae (Oligochaeta) from the west coast of Greenland. *Ann. zool. fenn.*, **7,** 199–209.
Patrick, F. M. and Loutit, M. (1976). Passage of metals in effluents through bacteria to higher organisms. *Wat. Res.*, **10,** 333–5.
Patrick, F. M. and Loutit, M. (1978). Passage of metals to freshwater fish from their food. *Wat. Res.*, **12,** 395–8.
Pfannkuche, O. (1977). Okologische und systematische Untersuchungen an naidomorphen Oligochaeten brackiger und Limnischer Biotope. PhD dissertation, Univ. Hamburg.
Pickavance, J. R. and Cook, D. G. (1971). *Tubifex newfei* n.sp. (Oligochaeta, Tubificidae) with a preliminary reappraisal of the genus. *Can. J. Zool.*, **49,** 249–54.
Pierantoni, U. (1902). Due nuovi generi di Oligocheti marini rinvenuti nel Golfo di Napoli. *Boll. Soc. Nat. Napoli*, **16,** 113–15.

Poddubnaja, T. L. (1965). Feeding of *Chaetogaster diaphanus* (Naididae, Oligochaeta) in the Rybinsk Reservoir. *Trudȳ, Inst. Biol. Vodokhran.*, **9,** 178.

Poddubnaja, T. L. (1980). Life cycles of mass species of Tubificidae (Oligochaeta). In *Aquatic oligochaete biology*, ed. R. O. Brinkhurst and D. G. Cook, pp. 175–84. Plenum Press, New York.

Potter D. W. B. and Learner, M. A. (1974). A study of the benthic macro-invertebrates of a shallow eutrophic reservoir in South Wales with emphasis on the Chironomidae (Diptera) their life histories and production. *Arch. Hydrobiol.*, **74,** 186–226.

Righi, G. and Kanner, E. (1979). Marine Oligochaeta (Tubificidae and Enchytraeidae) from the Caribbean Sea. *Studies on the fauna of Curaçao and other Caribbean islands*, **63,** 44–68.

Rofritz, D. J. (1977). Oligochaeta as a winter food source for the old squaw. *J. Wildl. Manage.*, **41,** 590–1.

Siebers, D. and Ehlers, U. (1978). Transintegumentary absorption of acidic amino acids in the oligochaete annelid *Enchytraeus albidus. Comp. Biochem. Physiol.*, **61,** 55–60.

Singer, R. (1978). Suction-feeding in *Aeolosoma* (Annelida). *Trans. Am. microsc. Soc.*, **97,** 105–11.

Smith, F. E. (1900). Notes on species of North American Oligochaeta III. *Bull. Ill. Lab. nat. Hist.*, **5,** 441–58.

Southern, R. (1909). Contributions toward a monograph of the British and Irish Oligochaeta. *Proc. R. Irish Acad.*, **27,** 119–82.

Southern, R. (1913). Clare Island Survey part 48, Oligochaeta. *Proc. R. Irish Acad.*, **31,** 1–14.

Sperber, C. (1948). A taxonomical study of the Naididae. *Zool Bidr. Uppsala*, **28,** 1–296.

Stephenson, J. (1911). On some littoral Oligochaeta of the Clyde. *Trans. R. Soc. Edin.*, **48,** 31–65.

Stephenson, J. (1912). On a new-species of *Branchiodrilus* and certain other aquatic Oligochaeta with remarks on cephalization in the Naididae. *Rec. Indian Mus.*, **7,** 219–41.

Stephenson, J. (1922). The Oligochaeta of the Oxford University Spitsbergen Expedition. *Proc. zool. Soc. Lond.*, **1922,** 1109–38.

Stephenson, J. (1932). Oligochaeta from Australia, North Carolina and other parts of the world. *Proc. zool. Soc. Lond.*, **1932,** 899–941.

Thorhauge, F. (1976). Growth and life cycle of *Potamothrix hammoniensis* (Tubificidae, Oligochaeta) in the profundal of eutrophic Lake Esrom. A field and laboratory study. *Arch. Hydrobiol.*, **78,** 71–85.

Timm, T. (1970). On the fauna of the Estonian Oligochaeta. *Pedobiologia,* **10,** 52–78.

Tynen, M. J. (1966). A new species of *Lumbricillus* with a revised check-list of the British Enchytraeidae (Oligochaeta). *J. mar. biol. Ass. U.K.*, **46,** 89–96.

Tynen, M. J. (1969). Littoral distribution of *Lumbricillus reynoldsoni* Backlund and other Enchytraeidae (Oligochaeta) in relation to salinity and other factors. *Oikos*, **20,** 41–53.

Tynen, M. J. (1972). The littoral Enchytraeidae (Oligochaeta) of Anglesey and the Menai Strait with notes on habitats. *J. nat. Hist.*, **6,** 21–9.

Tynen, M. J. and Nurminen, M. (1969). A key to the littoral Enchytraeidae (Oligochaeta). *Ann. zool. fenn.*, **6,** 150–5.

Wavre, M. and Brinkhurst, R. O. (1971). Interactions between some tubificid oligochaetes and bacteria found in the sediments of Toronto Harbour, Ontario. *J. Fish. Res. Bd. Can.*, **28,** 335–41.

Westheide, W. and Bunke, D. (1970). *Aeolosoma maritimum* nov. spec., die erste Salzwasserart, aus der Familie Aeolosomatidae (Annelida, Oligochaeta). *Helgoländer wiss. Meeresunters.*, **21,** 134–42.

Wharfe, RVM GR. (1977). The intertidal sediment habitats of the lower Medway estuary, Kent. *Environ. Pollut.*, **13,** 79–91.

Whitten, B. K. and Goodnight, C. J. (1969). The role of tubificid worms in the transfer of radioactive phosphorus in an aquatic ecosystem. In *Symposium on radio ecology*, ed. D. J. Nelson and F. C. Evans, pp. 270–7. CFSTT Natl. Bur. Stds, US Dept Comm., Springfield, Va.

Wood, L. W. and Chua, K. E. (1973). Glucose flux at the sediment–water interface of Toronto Harbour, Lake Ontario, with reference to pollution stress. *Can. J. Microbiol.*, **19,** 413–20.

Wood, L. W. and Chua, K. E. (1977). Method for studying amino-acid flux at the mud-water interface. *J. Great Lakes Res.*, **3,** 29–37.

Index of species